STIRLING WINGS

ALSO BY JONATHAN FALCONER

Heathrow
Stirling at War
abc UK Military Airfields (with W. Gandy)
Concorde (with A. Burney)
RAF Bomber Airfields of World War 2
Boeing 747
RAF Fighter Airfields of World War 2
What's Left of Brunel

STIRLING WINGS

Jonathan Falconer

FOREWORD BY GROUP CAPTAIN T.G. MAHADDIE
DSO, DFC, AFC & BAR, CzMC, CEng, FRAeS, RAF (Retd)

SUTTON PUBLISHING LIMITED

First published in the United Kingdom in 1995 by
Alan Sutton Publishing Limited, an imprint of Sutton Publishing Limited
Phoenix Mill · Thrupp · Stroud · Gloucestershire

Paperback edition first published 1997

Copyright © Jonathan Falconer, 1995

All rights reserved. No part of this publication may be reproduced, stored in a retrieval system, or transmitted, in any form, or by any means, electronic, mechanical, photocopying, recording or otherwise, without the prior permission of the publishers and copyright holders.

British Library Cataloguing in Publication Data
A catalogue record for this book is available from the British Library

ISBN 0-7509-1517-X

Author's note:
The counties referred to in this book are as they were at the outbreak of war in 1939.

Endpapers, front: Stirling Is of No 1651 HCU in close formation. W7459:O is being flown by WO Ashbaugh with Sam Curtis as his flight engineer (see p.125). (Rover Group) Back: Stirling IV, PK237, was the one-thousandth built by Short & Harland, Belfast, on 6 December 1944. It is likely that she was never used and was scrapped in 1945. (Shorts Plc Neg No: ST719)

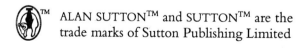 ALAN SUTTON™ and SUTTON™ are the trade marks of Sutton Publishing Limited

Typeset in 11/13 Bembo.
Typesetting and origination by
Sutton Publishing Limited.
Printed in Great Britain by
WBC Limited, Bridgend.

Contents

Foreword		vii
Acknowledgements		ix
Glossary of Terms and Abbreviations		xi
Introduction		xiii
	In the Beginning	1
1.	Early Days and Nights	8
2.	Giants in the Making	22
	An Inside View	33
3.	Over the Alps	37
4.	The Miracle Factory	43
5.	Down in Happy Valley	54
	Wings for Victory	67
6.	Life at Ground Level	74
	The Achilles Heel	80
7.	'A Piece of Cake!'	85
8.	For Gallantry	93
9.	The Old Man with the Scythe	100
10.	A Tale of Two Crews	108
11.	Most Secret	119
12.	The Dawn of Liberation	129
13.	Airbridge to Arnhem	136
14.	Death over Norway	146
	The Road to the Orient	157
15.	The Stirling Immortalized	161
	In Memoriam	168
Appendices		
I	Short Stirling I – Leading Particulars	173
II	Short Stirling – Principal Manufacturing Sites	176
III	Principal Stirling Airfields and Operators 1940–6	178
IV	Stirling Squadrons and Their Commanding Officers 1940–6	187
V	Named and Personalized Stirlings	202
	Bibliography	209
	Index	211

Dedication

To those Stirling crews who gave their lives in the cause of freedom

FOREWORD

Group Captain T.G. 'Hamish' Mahaddie DSO, DFC, AFC & Bar, CzMC, CEng, FRAeS, RAF (Retd)

I relish the opportunity to come to the defence of the Stirling, in spite of its operational shortcomings when compared to the Halifax and Lancaster. I find it most warming in the times since the end of the war to attend reunions of No 7 Squadron crews, where my keen interest and enthusiasm for the dear old Stirling is shared. True, we did suffer higher casualties than those above us, and from time to time we were party to damage by HE and incendiaries falling on us from above. In particular I remember a rear gunner in No 3 Group who, with his turret, was separated from his Stirling by a bomb from on high.

It took the genius of Don Bennett, the Pathfinder Force leader, to add nearly 8,000ft to the Stirling's meagre operating height of about 13,000ft by the simple expedient of reducing the quantity of .303in ammunition (based on what was usually brought home), cutting the fuel reserve of 22 per cent by a half, and further removing all the armour plating including the vast door, which measured ⅞ in thick, between the front cabin and the outer fuselage. I personally recall being at never less than 19,000ft on any target during my PFF tour, except on the original PFF first sortie to Flensburg.

I have pleasant memories of one special PFF captain, Wg Cdr Fraser Barron, who was known as the 'boy wing commander'. He flew two tours on Stirlings and was appointed commanding officer of No 7 Squadron in the spring of 1944 at the tender age of twenty-three. On his third tour, sadly he was killed when his Lancaster collided with the Deputy Master Bomber over the target of Le Mans, on one of the Transportation Plan raids prior to D-Day. Both he and his deputy were amongst the very best of Primary Markers and it was ironic that they should both die on a relatively easy target. This was a grim reminder that vast experience and outstanding gallantry were not passports to survival.

One of the most remarkable features of the Stirling was its ability to absorb battle damage. I remember in the first months of the PFF one of its founder members, Flg Off Trench, suffered a direct hit from HE which dislodged the port inner engine and left a large hole and a mass of wires and ancillary gear waving about in the slipstream. Nevertheless, Trench made it back to Oakington, and was immediately awarded the DSO, an honour I believe he shared with Len Cheshire as the only junior officers to receive such an honour at such a rank.

I would beg leave to present my personal experience, when my own Stirling, C-Charlie, got a squirt of 174 cannon shells from a nightfighter over the target of Cologne. It was clear that we must have been stalked by the Ju 88 prior to the actual strike, during which he used his upward-firing *Schräge Musik* twin cannon. While the aim was very accurate, oddly enough the shells ended up in the region of the bomb-bay and happily not in the area of the fuel tanks. Nonetheless, the attack immobilized the entire aircraft and wounded four crew. With hydraulics shattered and the guns useless, instrument panel disintegrated and master compass U/S, the navigator still made it to base on the astro compass.

In spite of the affection that crews had for the Stirling, it would be less than true not to stress how kindly they took to the conversion to the Lancaster. This was particularly so on No 7 Squadron, as I remember, when crews found that they were able to carry extra PFF stores further and higher than hitherto.

The Stirling had a short but effective life during the Harris offensive. Then despite being relegated from the first division it came into its own once more on the ill-fated Operation 'Market Garden' when it towed gliders. Indeed, the Stirling became a warm favourite with the Glider Pilot Regiment. It is small wonder, therefore, that Stirling crews testify to their undying faith in this fine old aircraft. Despite its limitations, it is remembered with pride and gratitude.

Group Captain T.G. Mahaddie
July 1995

Acknowledgements

I am indebted to the following individuals whose help so willingly given has been invaluable. Without them this book could not have been written:

Albert Albretsen and Jan Thygesen of Norway, for kind permission to use their account of the Stirling crashes in Augst-Agder county, Norway, in 1945; Mrs Connie Annis; Jack Atkins, wireless operator, No 149 Sqn; Gerry Blacklock DFC, pilot, No 7 Squadron; Peter Boggis DFC, pilot, No 15 Squadron; Chaz Bowyer; Leonard Brock, flight engineer, No 299 Sqn; Robin J. Brooks; Mrs Mary Brown; Bill Burns, flight engineer, No 90 Sqn; G. Carr, airframe fitter, 208 Staging Post, Pegu, Burma; Ernest Chappell, groundcrew, No 7 Sqn; David Collyer; Colin Cooper; Mrs Nora Curtis; the late G.E. 'Dickie' Dawe, pilot, No 1427 Flt, No 8 Ferry Pilots Pool, ATA; Bill Drinkwater; J. Earnshaw; Arthur Edgley, air gunner, No 15 Sqn; Mrs S. Ennis; Dennis Field, pilot, No 90 Sqn; Alan Gamble, wireless operator, No 620 Sqn; Bill Gibbs, ground staff, Nos 149 and 620 Sqns; Cal Gilbert; Ray Glass, pilot, Nos 214 Sqn and 1665 HCU; Tony Gray, instrument repairer, Nos 15 and 622 Sqns; Ray Griffiths, The Aircrew Association; J. Hardman, groundcrew, No 214 Sqn; Den Hardwick, pilot, Nos 149, 299 and 46 Sqns; Ted Harris; N. Harrison, groundcrew, No 570 Sqn; Bill Hotchkiss, wireless operator, Nos 218 Sqn and 1651 HCU; J. Hibbs, navigator, No 196 Sqn; Mrs Emily Howe, factory inspector, SEBRO; Michael Kidd, for his help with details about the Norwegian Stirling crashes in 1945; Hamish Mahaddie DSO, DFC, AFC, pilot, No 7 Sqn; Peter Jacques; Gil Marsh, pilot, No 622 Sqn; the late Len Marsh, fitter – aero engines, No 7 Sqn; Mrs Joyce Marsh; Jim Marshall, pilot, No 620 Sqn; K. Marshall, pilot, No 199 Sqn; Jim McIlhinney, navigator, No 218 Sqn; D. Mepham DFM; Ken Merrick; David Mitchell, navigator, No 149 Sqn; Jimmy Morris, flight engineer, No 218 Sqn; Arthur Old, stress department, Short Brothers Ltd; Ivan Pacey, leading hand electrician, SEBRO; Del Paddock, Wellesbourne Aviation Group; Simon Parry; Thomas Peacock, airframe fitter,

No 570 Sqn; Murray Peden DFC, pilot, Nos 214 Sqn and 1699 CU; Dr Alfred Price; Keith Prowd RAAF, pilot, No 196 Sqn; Maurice Rattigan, factory worker, Lea Francis Engineering, Coventry; F.M. Rawlings, engine fitter, No 6570 Servicing Echelon, Harwell; Miss Yvonne Reynolds; R. Smith, navigator, No 199 Sqn; J.H. Spiby, acceptance test pilot (RAF), Stradishall and Chedburgh; J. Stewart DFC, pilot, No 570 Sqn; Mike Stimson, wireless operator, No 196 Sqn; Alan Thomsett; R. Thorn; John Wheatley, Performance Testing Squadron, Aircraft & Armament Experimental Establishment, Boscombe Down; Mrs E.G. Wilkin; George Wright, wireless operator, No 622 Sqn; the late Vincent Wright, leader, the Stirlingaires Strict Tempo Dance Band.

I am grateful to my friend Group Captain Hamish Mahaddie for so kindly agreeing to write the foreword.

Acknowledgement is made to the following authors and publishers for kind permission to quote from copyright material: the Estate of H.E. Bates for a passage from *The World in Ripeness*; the late Air Commodore John Searby and HarperCollins for a passage from *The Everlasting Arms*; the late Dame Laura Knight and HarperCollins for a passage from *The Magic of a Line*.

The staff of the Public Record Office, Kew; Bath Reference Library; and the Commonwealth War Graves Commission, have been most helpful in the course of my research; so too have the following newspapers who kindly published my requests for help: *Basildon & Southend Echo*, *Cambridge Evening News*, *Coventry Evening Telegraph*, *East Anglian Daily Times*, *Leamington Evening Telegraph*, *Leicester Mercury*, *Liverpool Echo*, *Manchester Evening News* and *South Wales Echo*.

Thanks are also due to the following publications and organizations for permission to reproduce illustrations in their custody: Imperial War Museum, London; Military Aircraft Photographs; 199 Register; *Punch*; 75 Squadron Association; Shorts Plc, Belfast; Stirling Aircraft Association; Rover Car Group; *Aviation Photo News*/Brian Stainer.

Last, but not least, I must thank my wife, Annie, for her patience over the past four years. She can now have her dining-room back!

Glossary of Terms and Abbreviations

AA	anti-aircraft (gun)
A&AEE	Aeroplane & Armaments Experimental Establishment
ATA	Air Transport Auxiliary
AFC	Air Force Cross
B/A	Bomb-Aimer
Capt	Captain
CGM	Conspicuous Gallantry Medal
C-in-C	Commander-in-Chief
CO	Commanding Officer
DFC	Distinguished Flying Cross
DFM	Distinguished Flying Medal
DR	Dead Reckoning (a simple method of point-to-point navigation)
DSO	Distinguished Service Order
DZ	Drop Zone (for paratroops and air-dropped supplies)
ETA	Estimated Time of Arrival
F/E	Flight Engineer
Flg Off	Flying Officer
Flt Capt	Flight Captain (ATA)
Flt Lt	Flight Lieutenant
Flt Sgt	Flight Sergeant
ft	feet
FTR	Failed to Return
Gee	Radio navigation aid
HC	High Capacity (bomb)
HCU	Heavy Conversion Unit
HFF	Heavy Freight Flight
hrs	hours
i/c	in charge
in	inches
LAC	Leading Aircraftsman
LZ	Landing Zone (for gliders)
MAP	Ministry of Aircraft Production

Mk	Mark (eg, Stirling Mark I)
MO	Medical Officer
NCO	Non-Commissioned Officer
NZ	New Zealand
ORB	Operations Record Book
OTU	Operational Training Unit
Pitot Head	Air-pressure activated instrument for measuring aircraft's air speed
Plt Off	Pilot Officer
RAMC	Royal Army Medical Corps
RCAF	Royal Canadian Air Force
RNAS	Royal Naval Air Station
R/T	Radio Telephone or Radio Telephony
SAS	Special Air Service
SD	Special Duties
Sgt	Sergeant
SOC	Struck Off Charge
SOE	Special Operations Executive
Sqn Ldr	Squadron Leader
VC	Victoria Cross
Wg Cdr	Wing Commander
Window	Paper-backed metal foil strips dropped from bomber aircraft to cause interference on German ground radar screens
WO	Warrant Officer
Wop/AG	Wireless Operator/Air Gunner
W/T	Wireless Telegraphy

Introduction

For those who are new to the Stirling, one pilot once said that the best way to get used to the height of its cockpit above ground was to ride on the front seat on the top of a double-decker bus. Towering as it did some 22ft 9in above the tarmac, the view from the cockpit coupé was very different to that of a Wimpy or a Hampden, and its ground-handling characteristics were equally unique. Viewed from the ground, perched on its tall gangling undercarriage, the Stirling's great height was probably its most striking feature.

Of the eight heavy bomber types in service with the main combatant nations during the Second World War, the RAF's Stirling was unique in a number of vital respects: it was the tallest at 22ft 9in, the longest at 87ft 3.5in, and the slowest at a maximum speed of 260mph; at 99ft 1in it had the shortest wingspan, at 44,000lb the greatest empty weight, and the shortest range with a maximum payload, at 740 miles.[1] Of the three British 'heavies' it was the only one to be designed from the outset to take four engines, and the only one to have full dual control.

The Stirling was introduced into RAF squadron service in August 1940 and a total of 2,371 were eventually built and flown by the RAF. With the arrival of the Halifax and Lancaster in 1941–2 the Stirling quickly became the poor relation of the RAF's trio of heavies, principally for its lack-lustre performance.[2] Lack-lustre it may have been in the eyes of the Halifax and Lancaster crews but, as the saying goes, 'Beauty is in the eye of the beholder', and to the majority of Stirling aircrews their kite was undisputably the best.

Although 'officially' it was hard-pushed to reach more than 13,000ft with a full load, there are fully documented instances of individual Stirlings bombing the target from heights of 18–19,000ft with a 7,000lb bomb-load; others are recorded as reaching more than 21,000ft on unloaded test-climbs. Thanks to its solid construction the Stirling was able to absorb an amazing amount of battle damage and still make it home. Following a nightfighter attack one badly damaged Special Duties Stirling, on fire and flying on one engine, made it home across the North Sea with wounded crew on board; at 13,000ft over Berlin a Main Force Stirling was attacked by a Junkers Ju 88 and went into an uncontrollable power dive after its pilot was critically injured and slumped forwards across the controls. It took the combined strength of the wireless operator and navigator to level the aircraft out at 1,500ft, miraculously without it suffering any structural damage. With testimonials like these it comes as little surprise when former crews argue that, with the Stirling, it was a classic case of 'give a dog a bad name'.[3]

At the zenith of the Stirling's operational career with Bomber Command in 1943, just twelve squadrons were equipped with Short Brothers' mighty bomber before unacceptably high losses forced its relegation to second-line duties.[4] In its modified guise as the Mk IV, the Stirling fulfilled an important role as a paratroop transport and glider tug with No 38 Group until the war's end. In the opening months of peace a further modified variant, the Mk V, served briefly with the RAF as a long-haul freighter and passenger transport before it was superceded by better purpose-built aircraft like the Avro York. At the close of its RAF flying career in July 1946, 641 Stirlings of all marks had been lost through a combination of enemy action and accidents, representing more than a quarter of total Stirling production.

The publication in 1991 of my book *Stirling at War* prompted many who had not contributed to write and offer hitherto unpublished material for possible inclusion in a future volume. Long-forgotten photograph albums and flying log books were dusted down, documents and personal letters retrieved from writing bureaux, and memories rekindled of great – and some less hair-raising – deeds with the Stirling squadrons of the RAF. The results can be seen and read in the pages that follow. As with *Stirling at War*, *Stirling Wings* is not intended as a definitive service history of the aircraft, but rather a sequence of 'snapshots' covering the Stirling's operational service with the RAF during the Second World War. The words are those of the men and women themselves who built, flew or maintained the aircraft. The linking narrative is my own.

Jonathan Falconer
Bradford-on-Avon
April 1995

---

1. Comparisons made with Boeing B-17 Flying Fortress, Consolidated B-24 Liberator (USA); Avro Lancaster, Handley Page Halifax (UK); Petlyakov Pe-8 (USSR); Piaggio P108B (Italy); Heinkel He 177 (Germany).
2. Comparative production figures for the Stirling's contemporaries are: Wellington 11,461, Halifax 6,178, and Lancaster 7,374.
3. No 149 Squadron's Mk I, N6080, bombed Essen on 16/06/42 from 19,000ft, and reached 21,000ft on a test-climb on 25/06/42.
4. By comparison with the Avro Lancaster this is a tiny figure; at the war's end the Lanc equipped sixty out of Bomber Command's eighty heavy bomber squadrons.

INTRODUCTION

FIVE GENERATIONS OF THE STIRLING:
Mk I – 712 built. This example, N6090, served with the A&AEE, and Nos 7 and 15 Squadrons, before it was written off at Alconbury on 18 April 1942 when its undercarriage collapsed. (Author's collection)

Mk II – intended for construction in Canada, only two examples were ever built of this Wright Cyclone-engined variant. The second prototype, L3711, is pictured here at Boscombe Down in March 1942. (Imperial War Museum [IWM] MH5160)

Mk III – 1,047 built. EF411 completed at least sixty-nine ops with Nos 15 and 149 Squadrons before eventually passing to No 1653 Heavy Conversion Unit (HCU). (Author's collection)

Mk IV – 450 built. Newly completed Mk IVs await delivery to the squadrons from Shorts' airfield at Long Kesh, Northern Ireland, in September 1944. (Shorts Plc Neg No: ST641)

Mk V – 160 built. PJ943 photographed on a manufacturer's test-flight in March 1945, before passing to No 242 Squadron that same month. (Shorts Plc Neg No: ST711)

IN THE BEGINNING

'It was not a good-looking machine. Indeed compared with the shapely if somewhat rotund Wimpy it was rather ugly and badly proportioned.' *Gerry Blacklock, pilot, No 7 Squadron*

'The first sight of the Stirling's enormous size was rather daunting but actually flying it didn't turn out to be the frightening experience one had expected.' *Peter Boggis, pilot, No 15 Squadron*

While the Battle of Britain was gradually building to a crescendo in the early summer of 1940, the first Stirlings to leave the production lines of Short Brothers at Rochester were sent to Boscombe Down near Salisbury for performance evaluation by the RAF's Aircraft & Armament Experimental Establishment.

Gerry Blacklock, a sergeant pilot with a DFM and one tour of ops on Wellingtons with No 99 Squadron to his credit, was told to report to Boscombe in early July. Gerry and two others motored down to Salisbury on a beautiful summer's day, with maps to guide their path because all signposts had been removed to foil the Germans if they invaded. Nearing Salisbury, their first sight of Boscombe Down on the opposite side of the valley revealed an enormous gaunt-looking aeroplane, silhouetted against the skyline. Sure enough, the giant aeroplane was a Stirling:

'Instructions had been left for us to report to "B Per T" Flight the following morning, 4 July 1940. That unusually puzzling title meant, so we discovered, Bomber Performance Testing Flight. It was, however, a rather modest outfit: the flight commander, Sqn Ldr Collins, had one other pilot, Flt Lt Slee, and a Flt Sgt Thomas to look after the technical business. Our Stirling Development Flight in fact outnumbered our hosts. Our flight commander was Sqn Ldr Harris who had completed an operational tour with No 149 Squadron and had brought with him a skeleton crew of navigator, Sgt Austin, and rear gunner, Sgt Molyneux. There was my skeleton crew and Plt Off Reg Cox. A week or two later we were joined by Flt Lt (Saskatoon) Smith, a Canadian who had also completed a tour with No 149 Squadron. We also had half-a-dozen sergeant fitter 1s as a nucleus technical staff to learn the intricacies of the Stirling and to pass on their technical knowledge to our future groundcrews. In addition to our lot there were two civil airline pilots, Capt Messenger from Imperial Airways, and Capt Thomlinson from British European.

'That first day Sqn Ldr Collins gave us a 50-minute demonstration flight, but only Messenger and Thomlinson were given a chance to handle the controls and do circuits and landings. Cox, Treble and I watched from the platform behind their seats – as Reg Cox remarked, "like ruddy bridesmaids". [Messenger and Thomlinson now belonged to the Air Transport Auxiliary and would soon be delivering the aircraft from the factories to the RAF.] In fact, that was the way it was to be for the next three days. On 8 July, however, Collins checked out Reg Cox and me, an hour between us, and we were deemed fit for solo and, since there was no one to make further checks, we were automatically captains designate.'

The next notable event for the Stirling Development Flight was a visit from the 'Father of the Royal Air Force' himself, Marshal of the RAF Sir Hugh Trenchard. He informed them that the following week they were to become No 7 Squadron (a bomber unit with a distinguished history stretching back to May 1914, but which recently had been disbanded at Finningley in May 1940). Anything less like a squadron than the few aircrew and groundcrew mustered around the great old man would have been difficult to imagine.

'On 3 August I gathered my kit and crew into the only serviceable Stirling, N3640, and off we went to our new home at Leeming beside the Great North Road in Yorkshire. No 7 Squadron was officially formed on 1 August so when we arrived on the 3rd there was quite a large contingent of groundcrews already there to greet us. After a routine landing I saw that the aircraft was safely housed and then went off to find beds and settle in. Thinking that my work for the day was over, I was having tea in the Sergeants Mess when Reg Cox arrived in a bit of a sweat. General Sir John Dill, at that time Chief of the Imperial General Staff, was on the station and wished to be shown over the new giant bomber. The amount of brass congregated in the hangar was a little overpowering for a humble sergeant pilot but Sir John was very friendly and interested and, to my surprise, did not blanch at having to scramble through the spar in order to reach the cockpit, although that must have been quite difficult in riding boots and breeches. I was pleased his entourage did not follow.

'The following morning the CO, now Wg Cdr Harris, told me that from then on I was to be the captain of N3640 – the first four-engined bomber in Bomber Command.'

IN THE BEGINNING

The wing of the Stirling owed much structurally to the Sunderland flying boat, Shorts' other main warplane design of the Second World War. (*Aviation Photo News*/Brian Stainer)

HALF-MEASURE: In order to gain some experience of the Stirling's likely handling characteristics and appraise the suitability of various design features, Shorts built a half-scale prototype Stirling designated the Short S.31. (Military Aircraft Photographs [MAP] via R.J. Brooks)

TRIAL FLIER: This rare flying shot of the S.31 was probably taken in 1938 at Rochester during its early trials. (Shorts Plc Neg No: H1096D)

DOWN AND OUT: A binding wheel brake caused L7600, the first full-size Stirling prototype, to crash on landing at Rochester on 14 May 1939, after its maiden – and only – flight. (Shorts Plc Neg No: H1147E)

IN THE BEGINNING

THE SHORTS TEAM: Third from the left in the front row is Arthur Gouge, Shorts' chief designer and father of the Stirling, and next to him on the right is Oswald Short, chairman and managing director. John Lankester Parker, Shorts' chief test pilot, is at the right end of the front row. (Shorts Plc)

STRESS MANAGEMENT: Shorts' Stress Office personnel at Cuxton, Kent, with their head of department, R. Boorman, sixth from the left in the front row; Alfred Old is in the middle row, fifth from the left. (A. Old)

SECOND CHANCE: L7605 was the second Stirling prototype and is pictured here in December 1939. (Shorts Plc)

OFFICIAL RECOGNITION: This drawing, issued by the Ministry of Information in early 1940, was intended to represent the Short Stirling. Although it bore little resemblance to the Stirling, it was the first officially approved mention that such an aircraft existed. (Author's collection)

IN THE BEGINNING

FIRST OF THE MANY: An early production Stirling I makes a low pass over Shorts' airfield at Sydenham, Northern Ireland. (Shorts Plc Neg No: ST316)

CHAPTER ONE

Early Days and Nights

The date 10 February 1941 was significant in the history of No 7 Squadron as it marked its first operation in its second war. However, as operations went at that time there was nothing remarkable about it. Three aircraft (N3641, N3642 and N3644) were detailed to attack oil storage tanks at Rotterdam, so it was a trip of 3½ hours to a fairly easy target without a great deal of effective opposition. Gerry Blacklock recalls his involvement in this run-of-the-mill, but nevertheless significant, operation:

'The three senior captains naturally got the trip – Squadron Leaders J.M. Griffiths-Jones and P.W. Lynch-Blosse, and Flight Lieutenant G. Howard-Smith. I flew as second pilot to "Sask" Smith in N3641 and most of the other senior aircrew got in on the act somewhere. For example, we had the squadron navigation leader "Bunny" Austin, and the gunnery leader George Stock with us. Flying Officer Reg Cox went as second pilot with Griffiths-Jones and Flight Lieutenant Cruickshank (another ex-149 Squadron captain) even muscled in as navigator to Lynch-Blosse.

'On 24 February I did my second operation but this time I went as skipper with Reg Cox as second pilot and again "Bunny" Austin and George Stock, with my own regular crew members – Rossiter as wireless operator, Ashton as front gunner, and Taffy Price as flight engineer. We were bound for Brest where the German cruiser *Hipper* was in dry dock. The weather was a bit patchy but Bunny had no doubts about his sighting on the dock and we dropped our sixteen 500–pounders in three sticks from 10,000ft. Pat Lynch-Blosse had been less fortunate: having run into a snow storm on the outward journey he had jettisoned his bombs and landed at Boscombe Down.

'One evening late in March I had dropped into the Red Lion in Cambridge, an unusual occurrence because it was usually too crowded for my taste. But this time it was a fortuitous visit because Taffy Price was there and introduced me to his drinking companion, an air gunner with the name of Graham, who had just joined the squadron after a tour on Blenheims where he had recently earned a DFM. A word with George Stock the next morning

On 4 August 1940 Flg Off Gerry Blacklock became the captain of the very first four-engined monoplane heavy bomber in Bomber Command. (G. Blacklock)

N3641 was the second Stirling I to join No 7 Squadron. It later flew in the first Stirling raid of the war on 10 February 1941, target Rotterdam, skippered by Flt Lt G. Howard-Smith with Flg Off Gerry Blacklock as second pilot. (IWM CH3146)

secured Jock Graham for our rear turret. But still we lacked a regular second pilot and a navigator.'

Ernest Chappell was a pre-war RAF regular groundcrew, posted to Oakington shortly after the formation there of No 7 Squadron. He recalls the principal events of his first year on the station:

'In early January 1941 I was posted to Oakington to join a development flight of five yellow-bellied Stirling aircraft. These were being put through all sorts of tests by middle-aged civilian pilots and RAF aircrews, as well as groundcrew, all of whom were being trained to operate the aircraft. The new Stirlings were arriving from the factory at Rochester at the rate of two or three per week.

'From late 1941 parties of tradesmen from Oakington were posted to nearby No 3 Group aerodromes which were still flying Wellingtons to assist them in their transition to Stirlings, thus forming a full bomber group throughout East Anglia equipped with the new type.'

For most of February and March, No 7 Squadron had been flying from Newmarket next to the Rowley Mile racecourse, because Oakington's grass runways had many soft patches and thus were unsuitable for the operation of heavy aircraft. So it was from Newmarket that three Stirlings took off for Berlin on the night of 9 April. This would be the deepest penetration of enemy territory so far attempted by the Stirling as, apart from Cologne, all the previous targets had been coastal ones. It was also to be Gerry Blacklock's first visit to Berlin:

'Although N6005 had behaved well on her last two air-tests she decided to be temperamental that night. The port inner engine persisted in overheating and the only way to keep the oil temperature within bounds was, from time to time, to stop climbing and throttle back. Nevertheless, by the time we reached the Dutch/German border at Lingen we were at 18,000ft. The moon was full, there were no clouds and there was an impression of the land being washed in silver. However, we were given little time to admire the scenery before being held in a searchlight followed almost immediately by two or three more. At that time we had not heard of a "corkscrew" as an aerial evasion manoeuvre but I just did what amounted to much the same thing. I knew from one fighter affiliation exercise that the Stirling was pretty manoeuvrable and I now tested that quality to the full. But even so I was slightly shaken at one stage to find myself looking through the curved canopy above the windscreen directly into the searchlight bowls, then glancing quickly down I saw the artificial horizon full over – I reckoned it must have been a pretty tight turn!

'In the midst of these gyrations Stock yelled, "Fighter – break right!" and the fuselage was immediately filled with the clatter of machine-gun fire and the acrid stink of cordite, interrupted by a dull thump. Within seconds the attack ceased and the searchlights were doused. Stock said that the fighter was a Me 110 and he was sure that he had hit it, but we too had been hit. Rossiter reported that his W/T set had been shot up and the general consensus on the thump was that it had been in the starboard wing and we assumed one of the petrol tanks had been hit. But there was no indication that we were losing petrol nor were the flying characteristics impaired, so we resumed course for Berlin and set about regaining the height that we had lost.

'Unfortunately our aerial antics had not shaken the bugs out of No 2 engine which continued to overheat, and only by throttling back could it be returned to normal. So I decided to postpone my first visit to the capital and diverted to the secondary target at Emden docks, which we bombed from

N3638 was the fourth production Stirling I and joined No 7 Squadron in September 1940. It is pictured here at Belfast on 13 July 1940. (Shorts Plc Neg No: ST75A)

EARLY DAYS AND NIGHTS

16,000ft without further incident. Back at Newmarket we discovered that "Baggy" Sach [Flg Off J.F. Sach] had returned early with engine overheating and "Farmer" Pike [Flt Lt V.F.B. Pike in N6011] had not been heard of at all.[1]

'Next morning I went early to the dispersal hut and found my groundcrew chief, Sgt White. After exchanging civilities I went out to N6005 and there saw the cause of the thump – No 1 tank had been hit by an incendiary and being empty, apart from petrol vapour, had exploded. The tank was just like an empty sack and being positioned over the flap had made a large dent like a hip bath in the top surface. I felt glad that the flap had not stuck up last night.

'My next operation was to Kiel on 25 April but unfortunately the aircraft, N6013, developed a negative earth shortly after take-off so the bombs were dropped on the range at Berners Heath and we had an early night. The 27th was No 7 Squadron's first daylight operation with the Stirling using forecast cloudcover. Sqn Ldr Lynch-Blosse had that first trip but the forecast proved inaccurate as he ran out of cloud cover and brought his bombs back home. The next day Dennis Witt had better luck. The cloud stayed with him all the way to Emden where he laid his eighteen 500lb bombs across the town from 2,000ft before machine-gunning the docks.

'On the night of the 28th I went to Brest with Alan Naish, a civilian airways captain, and two nights later to Berlin with the CO, Bob Graham. We spent a long time over the city without being able to identify our aiming point due to haze and broken cloud below, and eventually brought our bombs home.'

The month of May brought visits to Brest, Bremen and Cologne (twice), and a new Stirling for Gerry Blacklock and his crew. N6022 was the new mount which quickly came to be regarded with considerable affection by all of the crew who took it on ten operations over the coming months, culminating with a trip to Magdeburg on 5/6 July. However, life for N6022 was destined to be brief: 'borrowed' from Gerry Blacklock by Dennis Witt on 15/16 July, the aircraft ran short of fuel returning from a raid on Hanover and was abandoned by its crew over East Anglia, who left the aircraft to crash at Newton Flotman, Norfolk.

Gerry Blacklock's first daylight operation in the Stirling was on 10 June when two aircraft from No 7 Squadron raided the docks at Emden in northern Germany. However, the met forecast for cloud cover proved to be a little optimistic as the only cloud to be seen from the Suffolk coast was away on the horizon, and it didn't look very thick at all:

'Being reluctant to turn back so soon, I persuaded myself that one could not really judge the quality of cloud from several miles away, so we pressed on for a closer look. However, we never reached the cloud layer, being picked up by

Silhouetted against the late afternoon sky, Stirlings of No 15 Squadron line up on the perimeter track at Wyton, ready for night ops, November 1941. In March of the same year the squadron had become the second to re-equip with the Stirling. (via P. Boggis)

two Me 109s when we were halfway across the sea, so it was down to sea level and turn for home.

'Jock Graham who had been our rear gunner for the past six weeks was now able to show his skill and coolness by directing me which way to turn as soon as the fighters had committed themselves to an attack. Thanks to him we could see their tracers zipping past our wing tip on a tangent to our turning circle. The only difficulty lay in translating his Scots burr!

'It all worked very well and must have been most frustrating for the German fighter pilots, which may have caused the leader to hold on too long on his second attack because Jock shot him down. The wing man evidently got the message for instead of continuing with quarter-attacks he took a long sweep out to our starboard and came in for a beam attack. When he turned in to attack I started to turn towards him and one of his shots bounced off our starboard inner exhaust shroud. Ashton managed to get in a burst from the front turret which persuaded him to call it a day and go home, which was just as well for us as he had put a bullet into the hydraulic system for the front turret. I suppose it had been an unusual start to a heavy bomber tour.'

Flg Off Peter Boggis brings No 15 Squadron's Stirling I, N6086, 'MacRobert's Reply', in to land at Wyton in November 1941. (via P. Boggis)

On 28 June the met officer once again forecast cloud cover over the North Sea and Germany, so three crews each from Nos 7 and 15 Squadrons were briefed for a sneak daylight attack on Bremerhaven. The No 7 Squadron aircraft captains detailed for this operation were Sqn Ldr Dickie Speare (N6020), Flg Off Gerry Blacklock (N3663) and Flt Lt Collins (N6007). Gerry Blacklock recalls:

'The story was the same as on the daylight to Emden on the 10th, with a thin layer of cloud in the distance. However, I thought that by staying fairly low and steering well to the north of the Frisian Islands we might reach the cloud undetected. We were well on the way when Jock reported another Stirling, about two miles behind us and on a track to the south of ours, being attacked by fighters. I turned back to give him some support but before we could reach him we ourselves were attacked by nine yellow-nosed Me 109s.

'This time we were flying the CO's aircraft, N3663, which had a mid-upper turret and Taffy Price had taken up position there as soon as fighters were mentioned, so we had six rearward guns. Again Jock's directions proved that a good bomber, properly guided, could be a match for a fighter – or even for nine fighters – because we were not even scratched during the attack while Jock notched up another Messerschmitt, and he and Taffy claimed two more probables. I suspect that their claims were probably more than justified because before we joined the other Stirling, flown by Flt Lt Collins, the fighters had broken off.

'I was unable to get any reply from Collins on the R/T but managed to edge him round until we were headed for home, then dropped back to formate on him. This would put us in a better position to stave off any further attacks if the fighters came back, and it appeared that "Q" had already suffered some damage in the attack before we got there. Having failed to make R/T contact we tried signalling with the Aldis lamp but with no more success. We got the impression that "Q's" starboard-outer engine had failed, although I don't remember that it was feathered, but for whatever reason his speed fell off to such an extent that it was difficult to maintain formation. Then he began to lose height and since we were only at about 100ft to start with it was not long before he hit the sea.

'When the sea spray cleared we could see that the fuselage had broken in two just behind the wing and the rear half had disappeared, and within a few seconds the front half too slid below the surface. I had Paddy drop a smoke flare so that we could orbit the position – I had a half-formed idea that we might be able to drop our dinghy if anyone surfaced. After circling for about ten minutes, with all eyes on the water, we saw nothing and set course sadly for home.'[2]

In November 1941 No 149 Squadron became the third RAF bomber squadron to re-equip with the Stirling. N6103, E-Easy, is pictured at Mildenhall soon after the squadron re-equipped with the type. (Author's collection)

Sqn Ldr Smithers and his No 149 Squadron crew pose for the camera at Mildenhall on a freezing 16 January 1942. Behind them towers W7455:B which later passed to No 1657 HCU and with

whom it was written off in September 1943 after being attacked by a German Me 410 intruder over Suffolk. (via Mrs N. Curtis)

The next daylight operation for Gerry Blacklock came on 23 July when reconnaissance aircraft brought news that the 38,000-ton German battlecruiser *Scharnhorst* had slipped out of the French port of Brest and headed south. No 7 Squadron was to have three aircraft (N6035, N6037 and W7434) stand-by until she was located, each loaded with three 2,000lb armour-piercing bombs. The *Scharnhorst*'s destination eventually turned out to be La Pallice which lies some 200 nautical miles beyond Brest on the Atlantic coast, near La Rochelle.

'I don't think I had anything to do with calculating the fuel load but I was surprised when Group Captain Adams (the wing commander was away) asked me if I was satisfied that the full tanks would be sufficient for the trip. I thought that he should have asked Reg Cox who was then a squadron leader and who was to lead the formation. I said that my only worry was getting that load off the ground – it was a hot day with only a slight wind. The upshot was that he arranged through Group for a reduced fuel load and for us to land at St Eval in Cornwall, instead of returning directly to Oakington.

'Daylight jobs always engendered some extra excitement but this one became a kind of gala occasion because the crews detailed were Reg Cox, Dennis Witt and myself – the three longest-serving captains in the squadron and all ex-apprentices. As usual, Reg had a word for it when he asked control for permission for "Freeman, Hardy and Willis" to take off! It was a lovely sunny evening as we flew low over the Channel, made a wide sweep round the Channel Islands and Brittany, and then the long haul south-east over the Bay of Biscay, ending in a rated climb to arrive over the target at 14,000ft where we broke formation and bombed individually.

'It might have been better if we had stayed in formation as Reg's bombs straddled the ship while ours were a bit wide. There was some heavy flak about and poor Rossiter lost his W/T set with a piece of shrapnel through it. By the time we had bombed the Me 109s had arrived, so it was down to sea level with all speed. We were chased down by two fighters until one was hit by Jock and broke away, and the other stayed back out of machine-gun range taking pot shots with his cannon, much to the annoyance of Taffy Price who was again in the dorsal turret to practise his gunnery. Our remaining "escorts" soon lost heart when we got down to the seas and rejoined formation. It had been amazingly clear over the target but after our swift descent it seemed to be no time at all before we were overtaken by darkness, then it was just a long, long journey back to St Eval by which time the excitement had gone and we were glad that we had no further to go that night.'

In the event, No 3 Group's Stirlings had caused very little damage to the

No 7 Squadron's Stirling I, N3706, climbs over the East Anglian countryside during January 1942. Later that year the aircraft failed to return from ops to Bremen and was ditched in the sea off Borkum on 30 June. Five of the crew including the pilot, Flt Sgt M. Bailey RCAF, were killed; three became prisoners of war. (IWM CH4782)

Scharnhorst but once night fell, thirty Whitleys of No 4 Group followed up with an attack on the docks at La Pallice, and the next day (the 24th) an unescorted daylight attack on the battlecruiser was made by fifteen Halifaxes from Nos 35 and 76 Squadrons. Although damage to the *Scharnhorst* was slight, the Germans decided it would be wise to move her north again to Brest under cover of darkness on 24/25 July, where defensive measures and repair facilities were better.

No more daylight bombing raids were flown by the two Stirling squadrons after the La Pallice operation, but the *Scharnhorst* story was far from over. After moving north to moorings in Norway, she was finally sunk more than two years later in a desperate engagement with Royal Navy warships at the Battle of North Cape off the north coast of Norway on Boxing Day 1943.

1.   Shot down by a nightfighter piloted by Feldwebel Karl-Heinz Scherfling of 7./NJG1. Aircraft crashed near Lingen, Germany, with one survivor who became a PoW.
2.   Shot down at 16.22hrs into the North Sea, 20 miles off Flamborough Head by Me 109s of I./JG52. No survivors.

CHAPTER TWO

Giants in the Making

In the months that followed the Stirling's first offensive operations with RAF Bomber Command in February 1941, it assumed something of a celebrity status in the press. It was the RAF's first four-engined bomber aircraft to enter service and its great size was something of a talking point when compared to its twin-engined predecessors like the Whitley and Wellington. Glowing reports of its exploits over Germany and occupied Europe soon appeared in the national newspapers, while columnists in specialist aviation magazines such as *The Aeroplane, Flight* and the ever-popular *Picture Post*, penned in-depth evaluations of the aircraft and its amazing capabilities. Officially sanctioned publications from the Ministry of Information were as quick to seize on the propaganda value of

An early setback to production were the Luftwaffe raids on the Stirling production lines at Rochester and Belfast during August 1940. This photograph shows the aftermath at Rochester following the raid of 15 August in which six Stirlings were damaged and production temporarily halted. (Shorts Plc via R.J. Brooks)

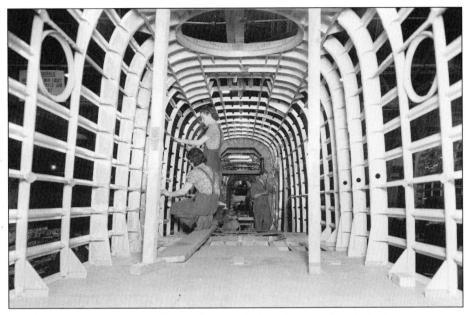

The Austin Motor Company at Longbridge, Birmingham, became the principal subcontractor to Shorts in the manufacture of the Stirling. Here a Stirling fuselage can be seen under construction at the Austin Shadow Factory, Longbridge, in the spring of 1943. (Rover Group)

such a large aircraft. In one such publication, *Bomber Command Continues* (HMSO, August 1942), the mass production of the Stirling was described:

'The method of producing Stirlings, and for that matter Halifaxes, Lancasters, and all other heavy bombers, is much the same as that used to produce Wellingtons; but a Stirling assembly plant at first sight offers a marked contrast. There is much more noise, more artificial light, and the hall itself is even larger. Look down on it from a balcony. It is so large that its further end is a blur under the fluorescent lighting which turns the faces of the workers green and their lips mauve. The noise of the machinery, too, produced mostly by the electrically driven tools, is intermittent but loud – loud enough to make it necessary to speak at the top of the voice and sometimes to shout.

'Here everything seems on a large scale. The fuselages, in their long rows, supported on bogies moving on a track which runs the whole length of the hall, look not unlike the bodies of Green Line buses. The wings, viewed from the end which will be attached to the fuselage, resemble dark, strutted tunnels into which a man may crawl with ease. Only the turrets appear the same size.

'This was one of the many shadow factories before September 1939. Since then the shadow has become a reality. Since the Stirling is a large aircraft with

Work on the complicated control runs and pipework inside the wings of what will become Stirling IIIs, is seen here being undertaken at Longbridge before the fuel tanks are fitted, 1943. (Rover Group)

four engines, weighing more than 30 tons, and containing more than 60,000 separate parts, the work of constructing and assembling takes longer than a smaller bomber like the Wellington.

'In order to ensure the maximum output and at the same time to avoid building complicated and costly jigs for which there are neither machine tools nor skilled labour available, large numbers of small, simple, easily made jigs are used, each jig or each group of jigs serving the main assembly section opposite which it is placed. The wide use of these sub-assembly jigs, as they are called, enables a new recruit in the factory to begin at once on an easy piece of work and thus to become proficient in a much shorter time than would be possible if the whole unit, fuselage or wing were built on a major jig. It is unnecessary and costly to make some dies and tools of steel in order to manufacture parts such as pipes and sheets of various shapes, whose material is thin-gauge light alloy. Many of these jigs and dies are therefore made of wood. Modifications in them can be made without delay.

GIANTS IN THE MAKING

'More complicated machines, such as the big cutters for shaping the spars, or the presses and boring machines, are so built that one operation is performed by each, and so placed that the operations take place one after another in a certain order. It is impossible to omit one operation, for a semi-finished part would not fit the next machine. Thus error is reduced to a minimum.

'Large though it be, the factory is not large enough to make and assemble all the parts which make up a Stirling. Many of these parts are manufactured elsewhere and sent in a finished or semi-finished condition to the Assembly Plant. Line production has been instituted and is indeed the rule in all aircraft factories. With Stirlings, both wings and fuselage are produced in lines.

'Before joining the line, the wing receives its first form on the building jig, where the leading and trailing edges and the engine nacelles are assembled. This

Control runs are installed inside an engine nacelle before the Bristol Hercules 'power-egg' is finally attached. The engine and its mounting were a complete and self-contained unit which could be taken from the airframe by undoing four bolts and disconnecting the auxiliary services and fuel lines on the firewall. This ease of maintenance was a great advantage to the groundcrews on active service. (Rover Group)

A scene inside the brightly lit Austin Shadow Factory early in 1941 with Stirling Mk I Series I fuselages under construction. (Rover Group)

RAF flight engineers receive familiarization training on the Hercules engine installation from Shorts staff at the Austin Shadow Factory in January 1941. (via Mrs N. Curtis)

wing shell is then carried by an overhead crane and placed on a specially constructed bogey running along a track which is the main assembly line and is parallel to a similar line on which the fuselages are being constructed. The wing then progresses stage by stage along the line till it reaches its fuselage. First the pipes and tanks are fitted into it, then the controls for the ailerons, then the landing and other lights, then the electrical equipment, then the engine controls. All the pipes in the various systems are pressure-tested after assembly.

'While the wings are thus being built, the construction of the fuselage is being carried out in much the same manner, though it is taken along the assembly line in sectional form. At each stage more and more items of equipment are incorporated into it, and finally the various sections of the fuselage – the nose, the forebody, the aftbody and the tail – are joined together. The turrets are then mounted, being dropped into position from overhead gantries. Then the testing gear is brought into operation, and everything in the fuselage and in the turrets is tested to make sure that it is in proper working order from the start. This testing at an early stage reduces the number of faults – or rather leads to their early discovery – and thus makes it

Four newly completed Stirling IVs and thirteen Mk Vs are pictured outside Short & Harland's Queen's Island factory in Belfast, early in 1945. Three Sunderland V flying boats and two Percival Proctor communications aircraft can also be seen. (Shorts Plc Neg No: J1075)

Flt Lt 'Dickie' Dawe RAF, in company with Flt Capt Joan Hughes (pilot) and Flt Eng Wally Drake, of No 8 Ferry Pilots Pool, Air Transport Auxiliary (ATA), at Sydenham in March 1943. 'Dickie' Dawe's task was to convert ATA ferry pilots on to the Stirling; the ATA crews' task was then to ferry new Stirlings from Short & Harland's factory in Northern Ireland to Bomber Command's No 3 Group airfields in East Anglia. (G.E. Dawe)

possible for a test flight to be carried out immediately the building process is ended. Before the wing is attached to the fuselage, complete tests are made of all electrical circuits, fuel tanks, control cables, etc.

'A schedule is attached to each main portion of the Stirling as it grows beneath the hands of the workers. This "Master Schedule" is a list of all parts and is added to as each part is inserted in the fuselage or wing. Thus at any moment the inspector or one of the supervisors, of whom there are three in the factory – one for the machine shops, another for the assembly lines, and a third for the press detail shops – can see by a glance at the schedule whether the aircraft details are progressing at the proper rhythm of speed through any particular stage. All stores are kept at the side of the main assembly tracks and the installations they contain are taken from them and inserted in the aircraft while it moves slowly past their position in the hall.

'After the fuselage and wings have been completed, the aircraft comes out of the main hall, and then passes into an adjacent block where it is painted, camouflaged and finally adjusted for flight. Then, black beneath and mottled above, it rolls slowly on to the tarmac of the aerodrome. There it is fuelled and taken into the air. When the test flight is over, the compass is swung, any adjustments which may have proved necessary are made, and it is then taken over by a Ferry Pilot. Stirlings thus built have been in action within forty-eight hours of leaving the works, and on one occasion one of them was over a target in Germany only twelve hours from the moment it touched down after its test flight.'

As already stated, many of the Stirling's constituent parts were manufactured elsewhere before being sent in a finished or semi-finished condition to the assembly plants. Dozens of sub-contractors in small factories and workshops scattered across the length and breadth of Great Britain produced a plethora of components. These ranged from complete bomb doors, tailplanes and throttle boxes, to electric motors, wheel tyres, aluminium extrusions and hydraulic remote controls. Each component, whether big or small, was a vital piece in the complicated jigsaw that, when fully assembled, became the mighty Stirling.

These many thousands of individual components were sent for final assembly at the principal manufacturing sites of Short Bros at Rochester, Kent, and Short & Harland at Belfast; and to the Austin Motor Co. Shadow Factory, Longbridge, and the Short Bros Shadow Factory at Swindon.

Before the Second World War Alfred Davies worked as a foreman at Carr's paper mill at Shirley on the southern outskirts of Birmingham. He tells us:

'When the war started the Ministry of Aircraft Production took the factory

Rochester-built Mk III, EF434, beats up Rochester airport during a maker's test flight in the late spring of 1943. (Shorts Plc Neg No: H1589A)

over to make parts for Stirlings and I was transferred to this work. At first I was making struts, then went on to making the bomb doors. I finished up being a chargehand on the rear wings.

'I was also in the Home Guard and during my two weeks of night work I helped look after local things; and during every two weeks of day work I helped the AA gunners at Swansborough Park at King's Heath, Birmingham, at night.'

In spite of the extensive damage inflicted on Coventry by the Luftwaffe's bombing of November 1940, the city still managed to contribute its fair share to the war effort. Maurice Rattigan was fourteen when he started work in September 1941 for Lea Francis Engineering Ltd in Much Park Street, in the heart of the city. He describes his experience as a factory worker.

'Prior to the war Lea Francis had made motor cars but when I joined them they were engaged on contract work for the MAP. They had differing varieties of work but their main contract was for petrol tank lids for Stirlings. For the same aircraft they also made throttle boxes, mudguards and oxygen bottle containers. The company made all the sheet metal components required – brackets, stringers, stiffeners etc – for the various products from the flat.

'There were four main petrol tanks in each wing and the tank lids varied in size from about 7ft x 5ft for the inboard down to about 5ft x 5ft for the outboard. At fourteen years of age, my first job was to de-burr the hundreds of pre-drilled holes in the outer skins at *3d* per skin.

'There were eight double-assembly jigs, each sufficient for two aircraft. As a lad I also "held up" on these jigs which involved putting the rivets in the holes and holding a dolly over them while the operator formed the head with a pneumatic hammer.

'I later worked on throttle boxes in the fitting shop but when I was sixteen I was moved to the tinsmith's shop where I was engaged on the manufacture of mudguards. Some of these came back damaged after service and I had to repair them.'

The story was the same in London where Ultra Electric at Acton was typical of the many factories in the capital engaged on contract work for the MAP. Ted Harris joined the drawing-office staff at Western Avenue, Acton, early in 1940 to work on the Stirling:

'The factory had already been stripped of the "Ultra" wirelesses and had a MAP order for 300 sets of rudders, bomb doors, elevators and instrument panels, followed up by another order of Breeze equipment (Blind Approach) for the same aircraft.

'The managing director was Edward Rosen who organized the transformation into a Shadow Factory, and obtained another follow-up contract of three hundred sets for replacement, so they said. Under his control little production time was lost through enemy bombing of the local Park Royal and Chase Estate areas.

'I was in the drawing office and was responsible for producing detail and assembly drawings from layout drawings supplied by Short Bros of Rochester. In the same office a young lady named Joyce looked after and redesigned instrument panels in a draughtswoman capacity, liaising with the Bristol Aero Company at Filton, Bristol. I later married her.'

AN INSIDE VIEW

The Stirling was a big aircraft from the outside but a very different story on the inside. There were two ways of entering the aircraft: a crew door at the back on the port side forward of the tail unit, and a hatch under the nose up at the front. The latter was rarely used because it required a very long ladder. With an aircrewman clutching his parachute, flying rations and flight bag, or a flight engineer's tool box, this combined weight made the ladder very unsteady and increased its tendency of sliding along the ground at the bottom. For the groundcrew, however, it was a useful and convenient means of access to the aircraft when topping up different services onboard by pipeline, or for loading ammunition belts. The rear door which opened inwards and to the right, required a small three-runged ladder in order to gain access, but it too was unstable if not a little springy.

With help from the groundcrew, aircrew would arrive inside the fuselage standing on a small step just inside the door. Down the fuselage to the right was the Elsan chemical toilet and another short ladder giving access, by way of a short tunnel, to the rear turret. Climbing a short distance down from the step you landed on a narrow aluminium walkway about 4ft wide. Turning to the left you continued for some 12ft up the length of the fuselage where you were confronted by a 4ft-high platform upon which flying kit could be deposited before scrambling up. This was actually the roof of the bomb bay which continued for another 42ft 7in forwards.

Three strides further on and you came across a small fixed three-rung ladder in the centre of the fuselage which gave access up into the mid-upper gun turret. After wriggling sideways-on around the turret mechanism and ladder – with full flying kit there was very little room to spare – you arrived in the space beneath the cabin roof escape hatch, also the home for ammunition storage boxes for the mid-upper turret. Passing through a small sliding door in the rear fuselage bulkhead you were confronted with the main spar.

Pushing your gear ahead you clambered over the spar and into the centre section where the crew rest bed was located. The name 'rest bed' was something of a misnomer since nobody had time to take a rest during the course of an operation. Its true purpose was to act as a bed for any crewmember who was wounded and it took up a fair amount of space. The mattress itself was covered with a plastic-like material to prevent blood from soaking through the ticking and soiling it. Overhead were two racks of oxygen bottles which were the main supply for the crew when flying above 10,000ft. Fitted in the roof on either side were seven levers for turning on and off the petrol supply to the fourteen fuel tanks. For the aircrew who needed to operate these

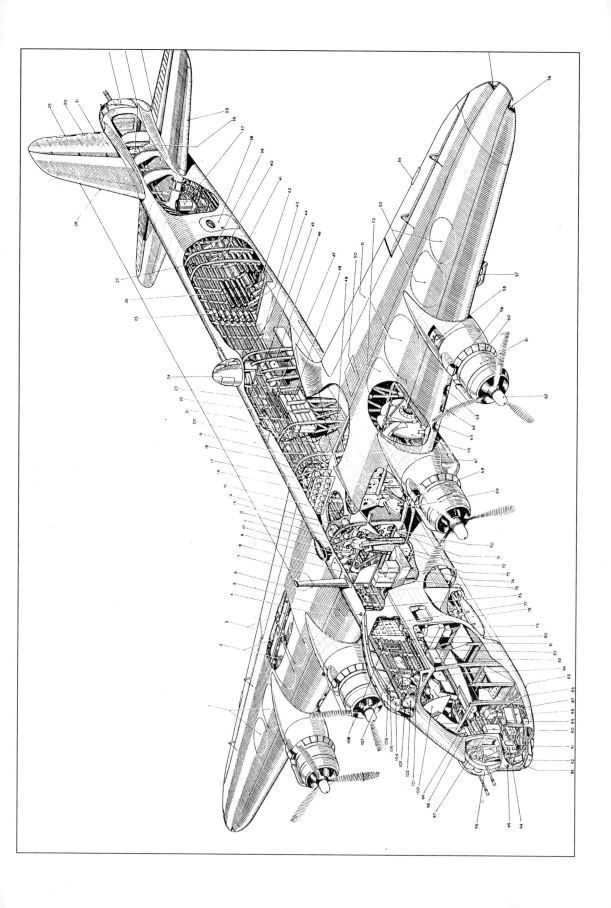

AN INSIDE VIEW

KEY

1. Fuel tanks (outboard inter-spar)
2. Fuel tanks (inboard inter-spar)
3. Fuel tanks (rear)
4. Upward identification lamp
5. Fuel tank (leading edge)
6. Aerial mast
7. Aerial mast de-icing equipment
8. Charging and distribution panel
9. D.F. loop
10. Heating system handwheel
11. Airframe de-icing control handwheel
12. Charging cable stowage
13. Oxygen bottles
14. Fuel cock controls
15. Bunk (rest station)
16. General service accumulators
17. Oxygen bottles
18. Airscrew anti-icing fluid tank
19. Observation dome (stowed)
20. Fire extinguisher
21. Fuselage heating pipe
22. Ground signalling strips
23. Escape ladder
24. Mid-upper turret (two Browning guns)
25. Flame floats or sea markers
26. Four reconnaissance flares
27. Fuel filler extension
28. Fin de-icing equipment
29. Rudder balance tab
30. Navigation lamp (tail)
31. Rudder trimming tab
32. Rear turret (four Browning guns)
33. Parachute stowage (rear turret)
34. Elevator trimming tab
35. Tail plane de-icing equipment
36. Bulkhead door
37. Elsan closet
38. Fuselage entrance door
39. Parachute stowage (flare launcher)
40. Parachute stowage (mid-gunner)
41. Four reconnaissance flares
42. Forced-landing flare chute
43. Reconnaissance flare chute
44. Maintenance trestles
45. Maintenance trestle struts
46. Tool roll
47. Ammunition conveyor (to rear turret)
48. Ammunition (rear turret)
49. Reserve ammunition
50. Dinghy stowage
51. Flap
52. Fuel tanks (inboard inter-spar)
53. Fuel tanks (outboard inter-spar)
54. Aileron trimming strip
55. Port formation-keeping lamp
56. Port navigation lamp
57. Twin landing lamps (retractable)
58. Oil tank
59. Carburettor air intake
60. Engine cooling gills
61. Engine { Stirling I – Hercules II or XI / Stirling II – Cyclone .G.R. 2600A5B
62. Constant-speed airscrew (feathering)
63. Oil cooling duct (outboard engine)
64. Cool air duct (cabin heating)
65. Oil cooler duct (inboard engine)
66. Port main undercarriage wheel
67. Undercarriage doors
68. Heating system boiler
69. Bomb in wing bomb cell
70. Carburettor air-intake & supercharger control handwheels
71. Ration tin
72. Fuel jettison control handwheels
73. Wireless operator's seat
74. Escape, and observation-dome hatch
75. Wireless equipment
76. Fire extinguisher
77. Engineer's instrument panel
78. Bomb in fuselage bomb cell
79. Navigator, 2nd pilot or bomb-aimer
80. Navigator's instrument panel
81. Compass
82. First aid outfit
83. Emergency tool kit
84. Automatic pilot control panel
85. Downward identification lamps (in floor)
86. Camera
87. Gyro Azimuth
88. Parachute stowage (bomb-aimer)
89. S.C.I. control panel
90. Ballast weight
91. Bomb-aimer's adjustable platform
92. Bomb-aimer's window
93. Automatic bomb sight mounting
94. Bomb-aimer's switch panel
95. Navigation lamp (nose)
96. Front turret (two Browning guns)
97. Drift sight stowage
98. Parachute stowage (front gunner)
99. Air bottles
100. Pilot's 1st instrument panel
101. 1st pilot
102. 2nd pilot's seat
103. Flap control panel
104. Cut-out control levers (Stirling I only)
105. Carburettor cock control levers
106. Rudder & elevator trimming tab gearbox
107. Compass mounting
108. Flying control locking gear stowage

Opposite: A cutaway illustration of the Short Stirling I. (Author's collection)

levers in an emergency if their flight engineer was incapacitated, many later remarked that it was a good thing the tank numbers were marked on the levers. Opposite the bed was the position from where the main undercarriage could be raised or lowered manually in the event of an electrical failure. The shaft came through the fuselage at this point, and the handle and other pieces needed for this operation were strapped beside it. Most flight engineers also kept their tool boxes tied securely to the floor in this area.

Climbing over the front spar you proceeded down a narrow gangway on the right hand side of the fuselage, passing the wireless operator's curtained-off cubby hole on the left and the flight engineer's seat and panel further on the right, just inside the heavy armour-plated doorway of the forward fuselage bulkhead. In the cockpit area that lay beyond – and where you could finally stand upright – could be found the navigator's seat and chart table on the left, curtained off as before.

At the very front were the two pilots' seats, positioned side by side on a raised platform, with a control column, wheel and rudder bar each. Between them was the throttle quadrant with throttle and constant speed controls, mixture levers, undercarriage selector, landing lamp adjuster, brake lever, and a comprehensive instrument panel before them. The first pilot's seat on the left hand side was fitted with a huge piece of downward-hingeing armour plate to protect his head and back. In the roof of the cockpit canopy were the main petrol cocks, slow-running cut-offs, tail trimming and rudder trimming cranks. In the V of the windscreen were the flap operation switch and indicator.

A hinged step in the centre-floor behind and below the pilots' platform gave access down two steps and forward into the bomb-aimer's compartment, where he lay prone on his stomach when sighting the target through the perspex clear vision panel in the floor of the nose. Access could also be gained from here up into the front gun turret situated in the extremity of the nose, above the bomb-aimer's position.

CHAPTER THREE

Over the Alps

After Italy's declaration of war on Britain and France on 10 June 1940, Bomber Command was quick to respond with long-range attacks on Turin and Genoa on 11/12 June. These were the first of many raids against Italian targets over the next three years which were largely aimed at undermining Italian morale. At first they were flown by Wellingtons and Whitleys, but the first Stirling raid on Italy was flown on the night of 10/11 September 1941, when thirteen Stirlings from Nos 7 and 15 Squadrons comprised part of the force of seventy-six aircraft despatched.

The four-engined Stirling was no more ideal for these long-range attacks than its twin-engined contemporaries, the Wellington and Whitley. Targets like Turin, Milan and La Spezia were at the extreme range of the Stirling, which meant that a full bomb-load had to be sacrificed for maximum fuel capacity. This payload meant that the Stirling could not climb its way over the towering switchback of Alpine peaks which stood in its way. The alternatives were to thread a perilous pathway through the mountain passes or to follow the western route across Lake Geneva, passing close to Mont Blanc itself.

While the rest of Europe was cloaked in the darkness of the blackout, it was a very

The Bristol Hercules engines of No 218 Squadron's Stirling I, Ha-K, are run up against the wheel-chocks on a ground-test. (*Aviation Photo News*/Brian Stainer)

different story in neutral Switzerland and across the Alps in Italy. Hamish Mahaddie, the famous Pathfinder pilot who completed a tour of ops on Stirlings with No 7 Squadron, recalls this bizarre contrast: 'On one of the occasions that we flew to Italy, we turned for Switzerland once south of Paris and there was the glory of Geneva – a bright smudge on the darkened horizon that was the city all lit up, with the lights actually being reflected on the lake – but several hundreds of miles away.'

The problems of topography and fatigue were sufficient in themselves to tax the flying skills of the Stirling crews and the performance of their aircraft, but added to this were other potential dangers. The loss of an engine through mechanical failure or enemy action meant a loss of height, and this could (and often did) spell disaster in mountainous terrain like the Alps. Unforecast strong winds, heavy rain and icing, and a myriad other adverse weather conditions, could lead to greatly increased fuel consumption, leaving an insufficient reserve to get the aircraft safely home to England.

Ray Glass, a second-tour pilot and flight commander with No 214 Squadron in late 1942, recalls: 'During November 1942 the raids on Turin were made more hazardous by severe icing conditions over the Alps, resulting in abortive sorties when aircraft were unable to clear the peaks.' Occasionally crews were forced to jettison their bomb-loads in desperate attempts to gain height over the mountains, forcing them to abort their sorties, about-turn and begin the long slog home across France, their outward journeys in vain.

It was well known that, compared to the Stirling, the Lancaster and Halifax made lighter work of crossing the Alps. The following observations were made by a Lancaster pilot named John Searby who, at the time, was a flight commander on No 106 Squadron stationed at Syerston, Notts. Returning from a raid on Turin during the night of 20/21 November 1942, he witnessed a homeward-bound Stirling re-crossing the Alpine peaks:

'On the way home that night I overtook a Stirling flying a little below me. This was in the region of the highest peaks. . . Apparently unharmed he was making steady progress and, no doubt looking forward to his bacon and eggs. As I came level he commenced a slow turn, dropping the nose until he was below the mountain top. In the bright moonlight I saw him make a feeble alteration of course which brought him clear of the shining peak but the nose was dropping even further until he plunged into a deep ravine to strike with a burst of flame followed immediately by a large explosion which lit up the icy walls – an enormous torrent of red light and then nothing more.

'It was a remarkable occurence because no parachutes were seen – no one attempted to get out though there was time. I wondered: oxygen starvation? Perhaps flak damage? Possibly. . . Amid the heat and flurry of a big raid one

OVER THE ALPS

Although breathtaking to look at, the Alps became the icy graveyard for many unlucky Stirling crews on long-haul trips to bomb targets in Italy. (Author's collection)

saw our bombers destroyed from time to time but this was almost outside one's consciousness – one willed it to be so because there wasn't time for reflection and every ounce of effort and concentration went into the task of getting the bombs on the target. This was different; in a peaceful setting, his job done, he went without fuss or bother to certain death.'[1]

Many years have passed since then. Snow and ice have covered the pitiful wreckage of this and many other aircraft which came to grief in this inhospitable terrain, encasing all in a glacial tomb. In the 1990s the ice began to melt, revealing the grisly debris of a number of wartime aircraft crashes.

Returning to the war years, the autumn of 1942 saw Sgt Jimmy Morris as a flight engineer with No 218 Squadron stationed at Downham Market, near Kings Lynn in Norfolk. On a raid to Genoa, he and his crew skippered by Flg Off 'Bickey' Bickerson witnessed the dramatic destruction of another bomber crew over the Alps. They themselves came close to not making it home either, with an insoluble mechanical problem which caused the engines of their Stirling to run in rich mixture and consume too much fuel:

'On 23 October 1942 we were briefed for a raid on the Italian target of Genoa. My friend from training days, Fred Hall, and his crew skippered by Plt Off R.A. Studd, were also detailed for this op. There were seven aircraft from No 218 Squadron involved. We took off in Stirling I R9196:G at 18.35hrs and all the rest got off within a few minutes of one another on what we knew was going to be a long, cold, hard slog. We crossed over the English Channel to France and continued towards the Alps, skirting around the north of Paris.

'About halfway across France I noticed we were using a lot more petrol on

Sgt Jimmy Morris, flight engineer, Flt Lt 'Bickey' Bickerson, pilot, and Flg Off Jim 'Jock' Traynor, navigator, pose for the camera beside their Stirling I, R9196:G, at Downham Market in March 1943. Note the ops tally of twenty-six bombs and ten sea mines.

Not long after this picture was taken Jock Traynor, a Scot from Shotts, Lanarkshire, was killed on operations aged twenty-three. Volunteering to stand in for another navigator who had gone sick, Jock flew with Flt Lt G.F. Berridge and crew on a minelaying sortie to the Heligoland and Elbe areas on 28/29 April. At 00.56hrs on 29 April his Stirling, BF515:N, was shot down by a nightfighter over eastern Denmark, crashing at the village of Taagerup. The crew of seven were killed, five being so badly mutilated in the crash that they were buried in a communal grave little more than a mile away in Reerslev churchyard. (J. Morris)

the starboard two engines than the port two. We had fourteen tanks on the Stirling – 2,240 gallons in all – and on a long trip like this one you could not afford to waste any. I did some checks with the skipper and we found out that the starboard engines were in full rich mixture which meant they were consuming far more petrol than they should. We had two mixture control levers, one on either side of the throttle box. This was a hydraulically controlled system (Exactor) and if it failed the two engines working off that system automatically went into rich mixture. This is what had happened. There was nothing I could do to alter the situation so we carried on with fingers crossed.

'We started to climb over the Alps which peak at about 16,000ft but there was no chance of getting the Stirling up to that height with the load we had on board, so it was a case of picking our way around the mountain tops at about 12,000ft. Now and again we would sight another aircraft doing the same. It was while weaving our way around the mountain peaks that our wireless operator, Sandy, said he was picking up somebody in trouble over the radio. Shortly after this we saw some flares being dropped which lit up the mountain tops and we could just see what looked like a Stirling flying above the glow of the flares until at last they burned out.

'We were still looking in the general direction where the flares had been when the whole of the mountain top was lit up by a blinding explosion and a great fireball rolled down the mountainside. This was the fiery end of one of the crews that had started out on the same op as us. Our navigator, Jack Traynor, made a note of the time and position of the explosion in his log and we carried on to the target. To this day I can still see the top of that Alpine mountain – the shape of it was indelibly stamped into my mind. That crew was so near to making it through the mountains.

'Once we had arrived on the Italian side of the Alps in bright moonlight we could see the target ahead of us as we made our way across the valleys. We saw a couple of elderly Italian fighters, one a biplane, but they didn't engage us. We bombed the target and started back for base.

'There followed a very steep climb to get over the top of the Alps and then an equally steep drop down into France to try and save some of our fuel as by this time it was getting quite low with the distance we still had to cover. I had a talk to Bickey the skipper and he decided to make for Manston which was the nearest English base to the French coast. Crossing the Channel I tried to drain as much fuel as I could from the tanks that had already been used and a couple of times the engines on one side cut out. I might add that the crew were not pleased at this and there was much muttering over the intercom. We made it to Manston at 01.40hrs – just! As we taxied to the dispersal point the engines gave a final cough and that was that. We flew back to Downham Market next morning to learn on our arrival that Fred my good mate, along with his crew, was missing. You accepted this situation on an operational station and just hoped that the missing crew would turn up safe, although most times they never did.'

In common with the four other Stirling squadrons operating that night (Nos 7, 15, 149, 214), No 218 Squadron's aircraft suffered their fair share of fuel shortages and engine failures. The squadron sustained the only Stirling loss of the whole operation from a total of fifty-one aircraft despatched, when Stirling I, R9184:U, captained by Plt Off R.A. Studd crashed into the sea off Dieppe, France. There were no survivors from its eight-man crew.

Every one of the remaining six crews from the squadron experienced problems of one sort or another that night:

W7614:J, captained by the squadron CO, Wg Cdr O.A. Morris, was attacked by a nightfighter north-east of Paris. The gunners opened fire but made no claims of damage against their attacker. The crew completed the operation and returned home to base at 03.00hrs.

Flt Sgt D.W. Thomson and his crew flying in W7612:T had a hair-raising time when their port-outer engine cut out several times as they crossed the Alps. They were forced to abandon the sortie and turned for home after jettisoning their bomb-load to lighten the aircraft, arriving back at base at 01.05hrs.

Flt Sgt J. Wood flying BF343:M also aborted, with port-inner engine trouble.

Sgt C. Jerromes flying BF346:S failed to make it further than France before he too was forced to return home on three engines after the starboard-inner had cut crossing the English coast.

Sgt L.T. Richards and his crew flying BF375:O almost lost out against French coastal flak batteries which nevertheless succeeded in holing several of their petrol tanks, forcing them to jettison their bombs in the Channel and head for home.[2]

(As to the ultimate fates of the individual aircraft themselves, R9196 was SOC 14/01/45 after a long spell as an HCU trainer; W7614 FTR Fallersleben, 18/12/42; short of fuel, W7612 crashed at Tangmere returning from a leaflet raid, 09/11/42; skippered by Flt Sgt G.A. Parkinson, BF343 crashed at Dieppe returning from a raid against Stuttgart 12/03/43; BF375 survived the war to fall to the scrap-man's torch; BF346 passed to No 90 Squadron and skippered by Sqn Ldr R.S. May FTR 28/04/43 from a gardening sortie to the Langelands Belt, Denmark, probably the victim of a nightfighter.)

1. John Searby, ed. Martin Middlebrook, *The Everlasting Arms* (William Kimber, 1988), pp. 44–5
2. PRO AIR 27 1350

CHAPTER FOUR

The Miracle Factory

With the Stirling squadrons of No 3 Group being centred in East Anglia, Cambridge was the ideal place in which to locate a repair organization to deal with damaged aircraft. Once established, this organization became widely known by its acronym SEBRO, which stood for Short Brothers Repair Organization.

At its peak SEBRO employed some 4,500 people at its Madingley Road factory, 2 miles west of Cambridge, and at RAF Station Bourn a further five miles to the west. The organization had the important task of repairing accident and battle-damaged Stirlings and, where possible, returning them to service to continue the fight against Nazi Germany.

Initially, SEBRO had used offices at King's College, Cambridge and hangarage at RAF Wyton from the end of 1940, but the quickening pace of the RAF's bomber offensive led to a corresponding increase in repair work for SEBRO. The result was the removal of its operations to a purpose-built wartime production facility, the first phase of which was completed in 1941, occupying a large site running south for about ¾ mile from its main entrance on the Madingley Road (now the A1303, originally part of the old A45), then turning east to Hangars Nos 4–7, built in 1942.

The first phase of the construction programme comprised Hangar 1 and interconnecting Hangars 2 and 3, with adjoining staff administration offices and a canteen building. Repair work by the initial workforce was concentrated in these hangars until Phase 2 of the construction of the main production hangars was completed during 1942. These were interconnecting Hangars 4 and 5, and double-length interconnecting Hangars 6 and 7, situated on the eastern side of the site. The workload continued to increase rapidly so Hangars 6 and 7 were quickly extended shortly after their initial completion to triple-length capacity. This enabled them to house eighteen Stirling fuselages side-by-side down the full length of Hangar 6, with port and starboard mainplanes lined up along both sides of Hangar 7.

Hangars 6 and 7 were now acting as the main production areas so the other hangars were re-allocated to factory support functions with No 1 becoming the main supplies store, Nos 2 and 3 the paint-spraying and finishing section, and Nos 4 and 5 for breaking up and parts salvage of Category 'E' scrap aircraft.

The men and women who made up the SEBRO workforce came from many areas of Britain and from as many different walks of life. It proved to be a mix of staff which brought together many skills and as much enthusiasm to the job of getting damaged Stirlings back into the skies. Management and senior technical staff came from the parent company of Shorts at Rochester. Other workers were directed to SEBRO by the Ministry of Labour while some were recruited directly by the company itself. A number of RAF personnel were also seconded under civilian status owing to the shortage of skilled people for the aircraft repair industry.

The initial grading of battle- and accident-damaged Stirlings was done at the parent airfield or wherever the aircraft had crashed. In the case of minor damage, Category 'A', the RAF dealt with the repairs itself but with aircraft requiring major repair work, radios, loose instruments and any secret equipment was removed before the wings and airframe were dismantled for transportation to SEBRO. Damaged but airworthy Stirlings were flown to Bourn for damage classification: those classified Category 'Ac' (damage repairable beyond unit capacity) were repaired at Bourn; Category 'B' (damage beyond repair on site, but repairable at a Maintenance Unit or at a contractor's works), together with Category 'C' and 'D' aircraft, were dismantled – either at their home base or at Bourn – and carried to the Madingley Road factory on Queen Mary low-loaders for attention. On arrival at SEBRO, damaged but repairable Stirlings were brought into the Category 'B' hangar while those beyond repair were taken to the Category 'E' hangars (Nos 4 and 5) where they were stripped of all instruments and working parts, some of which were returned to the Stirling production lines at Rochester, Birmingham, Belfast and Swindon for re-use.

Repairable fuselages were examined and received an assessment of the time considered necessary to complete the repairs and a target date set for delivery back to the RAF. Any unexpected setback in the work would mean that some people could be asked to work all night, having just done a 12–hour day, in order to meet the deadline. Once repairs had been completed, the aircraft were re-assembled at Bourn airfield and flight-tested.

Joining the SEBRO organization in 1941, Ivan Pacey was employed for the duration of the war as a leading hand electrician at Madingley Road. With so many of the Stirling's primary systems electrically operated, it was a demanding job, as he recalls:

> 'Early mornings and evenings saw a stream of people *en route* to or from the factory, on buses and cycling in the quiet locality of the university. Apart from access at the main gate there was an additional pedestrian and cycle access at

Ivan Pacey, leading hand electrician at SEBRO, pictured in 1943. (I. Pacey)

the far end of the factory site opening on to a public footpath which ran from the outskirts of Cambridge to the nearby village of Coton. This entrance was used by hundreds of us who worked in the main production hangars.

'When the factory first started in 1941 there was a small workforce on single dayshift working, with male employees working a considerable number of overtime hours. It was not unusual on a rush-job for a section to do a "Ghoster" – work all day, all night and continue through the next day. There was a typical strict prewar working regime without any mid-morning or afternoon tea breaks. With the completion of the main production hangars and the build-up of the workforce, two-shift working was introduced on a 12-hour day/12-hour night basis and on a monthly rota. With maximum production effort underway, working routines now included the much-needed tea breaks which, in addition to the canteen meals, helped to sustain workers through the long hours, thanks to the efforts of the canteen staff who contributed in no small way to the production effort.

'Stirlings needing major repairs were dismantled for transport to the factory with the mainplanes, less engines, and the fuselages going to different sections for repair under the supervision of the senior foreman of the section. Albert Welch had responsibility for mainplane repairs and Bill Brown and Bob Gedge were in charge of the two fuselage sections which were split between several chargehand supervisors.

'In the initial supervisory arrangements, responsibility and control of all the electrical work came under the chief electrical foreman Bob Nesbit, a demanding and energetic character, whom I understand had been very much involved with the Blackpool Illuminations in prewar days. But with the production organization established, our section was split up and transferred to work under three section foremen. Bob Nesbit took on a managerial role in a group liaison and technical service capacity. He was responsible for the design and introduction of our circuit simulator consoles for checking the fuselage electrical circuits. He also organized a course at the Ministry of Labour's Letchworth Training Centre to train the female works and inspection employees in the electrical sections on the Stirling electrical system.

'On hearing about the course I jokingly said to my chargehand Bert Leake that I wouldn't mind going on it. However, the remark was taken seriously by Bob Nesbit who decided I could be useful by helping to set up the demonstration mock-up of the Stirling wiring. And so I found myself at Letchworth for several weeks with twelve young ladies – fortunately their feminine wiles left me with no permanent scars. The success of the training course contributed to the skills and effort shown by the women of the SEBRO workforce.

'With the Stirling's design using mainly electrically operated primary equipment, ie: undercarriage, bomb-doors, flaps, plus all the other electrical services in an aircraft of its size, the servicing and repair of the electrical system called for considerable technical and practical effort. It was absolutely essential to ensure that when a fuselage and mainplanes left the factory for assembly at Bourn airfield, the electrical system was in sound order and the circuit wiring fully checked. The considerable array of wiring to various equipment and the multi-cable conduit runs to interconnecting junction boxes, instrument panels and the wing root connectors posed a formidable challenge in practical terms.

'Although there was a lot of space in the fuselage there was often quite a scramble when several operatives were all trying to work in the same section and we frequently had to withstand the ear-bashing noise of riveters hammering away as we worked on diagnosing and repairing faults. One

THE MIRACLE FACTORY

sometimes needed to be a contortionist to get at some parts and carry out repairs. On mainplanes repair work needed to be cleared before the petrol tanks were refitted and most importantly before the chap nicknamed "Treacle" – owing to the oily condition of his boiler suit – could refit the oil tanks back into the engine nacelles, making access to the wiring and particularly the propeller-feathering solenoids difficult. Another problem was posed when the combined dampness of the Cambridge fens and humid conditions caused the final inspection on the wiring insulation test to drop below the acceptable one megaOhm reading. The only solution was to pray for a clear fresh breeze in the early morning when the reading would usually rise to the required inspection standard. There was a desperate rush to waylay an inspector before the heat of the day returned.

'The electrical inspection testing was the final factory repairs operation so any problems at this stage caused some gnashing of teeth. On one such occasion following inspection of the conduit and junction box terminals, as I was refitting the conduits I spotted that a pin on Junction Box No 1 was partly corroded through. JB No 1 was a mass of wires and terminations and, positioned by the pilot's seat at floor level, a nightmarish job to replace. Thankfully the rush-job was completed by our night shift team and a potential future fault averted.

'We also experienced the mysterious and the dramatic, like the time when we found the detonator ring circuit on one of the fuselages had been cut in several places. In the event of a crash-landing in enemy occupied territory, the circuit ensured destruction of onboard secret electronic equipment to prevent it from falling into enemy hands. The mystery of who or why was never solved. The 'dramatic' was the discovery of a couple of bombs still in the racks of a mainplane being broken up in the Category E hangar. Fortunately some local RAF armourers were able to deal with the problem before we had more scrap bits flying around than we needed.

'As the war advanced, SEBRO was involved in converting some Stirlings for the role of paratroop dropping and modification for use as transport aircraft. Being solely a factory with no airfield, regrettably we saw few of our Stirlings in flight other than those on operations from surrounding aerodromes. But I recall one special occasion when we were privileged to see one of our specially prepared versions in an all-silver finish flown from Bourn over the factory. We were treated to a brief aerobatic display demonstrating the bird-like manoeuvrability of the Stirling, the silver finish accentuating its clean lines. Quite a grand and satisfying sight.

'Like most wartime factories there was a good spirit of cooperation and work effort with the serious side of work punctuated by lighter moments of

A Stirling Mk I Srs III, repaired and ready for the attentions of the paint shop. The pungent smell of cellulose dope seemed to permeate almost every area of the SEBRO factory site. (Author's collection)

jovial banter. On many occasions we relaxed during the canteen break to the piano music of Charlie Bull, one of the chaps from the pipe-fitting section who was also the pianist in the 'Stirlingaires' dance band which had been formed at the factory. The canteen was also the venue for one of the "Workers' Playtime" wireless broadcasts when the principal artiste was the popular singer Dorothy Carless. We were also visited by the Minister for Aircraft Production – Sir Stafford Cripps, speaking to us of the importance of our work to the war effort.

'Much of the credit for the organization and drive at SEBRO must go to the Works Manager, George Gedge, a strong and popular character whose personal approach within the works brought a quality of leadership to the efforts of the workforce. It is also fitting to record that the skills and dedication applied to the work on Stirlings by all the people in the SEBRO group was reflected by a sense of pride in being part of the Stirling's wartime contribution to the role of the RAF.'

Such were the exigencies of wartime that many women were drafted by the Ministry of Labour to take on men's jobs in the factories, thus freeing the men to join the forces. Emily Hayward was one of these women and she joined the SEBRO workforce in January 1942 after finishing her trade training at the Ministry of Labour's Letchworth Training Centre. She recalls:

'At first there were only two other women besides myself using the staff canteen and the three of us, feeling rather conspicuous, lunched together every day. Mrs Gilman, an older worldly-wise arty lady from Hampstead was put in charge of the View Room where girls sat in rows examining nuts and bolts salvaged from crashed Stirlings to see if they could be used again. Mrs Faith Nesbit (who was Lady Faith, a daughter of the Earl of Sandwich) was responsible for arranging lodgings for all the workers arriving daily from outside the area. I was given a little department of my own, known as Appendix "A". The items for which I was responsible were listed in a big book, each with a reference number. Some of the equipment was loose, such as Aldis lamps, inflatable dinghies, handles for winding down flaps, bomb-doors or the undercarriage when electrical systems failed; and some things were fixtures. Some were supplied and fitted by us, the contractors, while some were supplied by us and fitted by the RAF. There were code letters and numbers to identify all of this and needless to say there were hundreds of items in a four-engined bomber like the Stirling.

'At first there was not much work coming in so I used the time to pore over the book, familiarizing myself with all this equipment. What, for instance, was a Rear Gear Release Unit? I soon memorized most of the reference numbers, thus saving time in searching the book. Foremen and chargehands would seek me out for the information to save their own time. It was quicker to ask Lucy – the nickname for Loose Equipment!

'As soon as a crashed fuselage came in I had to go and list all missing and damaged items, then order replacements. The RAF usually cleaned the worst of the mess from the fuselages before they came to us, but as bombing missions increased so too did the crash-landings on return. The undercarriages were almost always damaged and there were many belly landings. Often we had to deal with the fuselage in the same state as it had come home, with the smell of blood and urine – and, on one occasion, a severed finger on the navigator's table. The pilot's seat had armour plating to protect the back and sometimes the seat had been shot through from below. The horror of war was brought home to us every day.

'We were working long hours, 7am to 7pm. Although there was no regular nightshift, there were often groups of workers having to stay and finish a fuselage on time. When it was ready I had to go in to ascertain that all my equipment was there. So, as others were clocking out at 8pm, the boss could say, "No 9109 is almost ready, will you stay on?". I would find myself starting again having done a full day's work. It was known as "doing a ghoster". There were moments when I felt that I was winning the war singlehanded.

'Working on Sundays meant being called out to an aerodrome, where the repaired Stirling had been delivered to the RAF before the works inspectors had passed it. However, any minor faults could be dealt with on the aerodrome. A jeep would be made available to take us, usually an inspector from the Aeronautical Inspection Directorate (AID) and myself. There were occasions when a car was put on just for me, making me feel like a VIP.

'Sunday, no other women working, the Stirling is teeming with RAF personnel preparing it for operations. No entry at the main door near the tail so I try the escape hatch. It is very high off the ground but the hatch is open and underneath stands a metal structure of scaffolding, dripping with oil. Even my rope-soled shoes would not grip on this. Note-pad between my teeth I climb, coat flapping in the wind. Having pulled myself up through the forward escape hatch with my eyes firmly shut – I have no head for heights – I start work, beginning with the front gun turret and working through the fuselage to finish in the rear turret. It would not have mattered too much if I had missed any of my items, the RAF ground staff would make sure that everything was there for the crew. But the items listed on our chart had to carry my stamp, SRC 42, and I could not stamp without looking personally.

'The work was increasing and it became necessary to take on more people to do Appendix "A" and this meant that I would be staying on in the office in charge of other workers. I was given the alternative of teaching another girl the work I had been doing to allow me to be transferred to a completely different job – testing electrical systems. This sounded interesting and I soon found myself once more at Letchworth Training Centre with a small group of girls learning about electricity.

'The lecturer was one of the design team from Short's who knew in detail about the Stirling's systems. We were taught only DC as the Stirling had only 12v battery power. For three months we went around muttering Ohm's Law and making the obvious jokes about magnetic fields being the place for lovers who offered no resistance. On inspection we had to know the circuit and test it, a different skill.

'It took two people to test the system when it was ready for inspection. Many of the cables carried multiple wires which, if connected to the wrong terminal, could create chaos – switch on navigation lights and find yourself opening the bomb-doors. With these sorts of consequences testing could be hilarious.

'All wiring to the wings was carried as far as the centre section of the fuselage. An eleborate construction of lights and switches which we called "the organ" was plugged on to the outside of the fuselage as a mock-up of the systems in the wings. The wings, of course, carried starter motors, navigation lights, bomb-release gear, flap motors, undercarriage – in fact almost everything of importance.

'Complete with headphones and intercom, one sat in the pilot's seat while the other manipulated the organ as necessary, pressing solenoid switches to bring in heavy duty cable. Flaps out, undercarriage down, landing lights on – we felt that we would have known how to fly. It was easy we decided, just switch over to George, the automatic pilot, when the course had been set.

'The mid-upper turret was a special case because it could turn full-circle, making it impossible to use wiring for electricity supply. The supply was conveyed by slip-rings which, of course, had to make perfect contact throughout 360 degrees. To test this contact, one inspector sat in the turret turning it with a handle, the other stood down in the fuselage, both using intercom headphones. It was vital for us to talk non-stop as the turret was turned in order to detect any gaps in the slip-rings. It can be difficult to talk to order but luckily my partner Ruth and I were poetry lovers and, being Cambridge, it had to be Rupert Brooke. Other workers in the aircraft would be somewhat surprised to hear one of us declaiming "The Soldier" all without punctuation.'

Despite working towards the common goals of winning the war and defeating Naziism, labour relations at SEBRO were not all they could have been. Emily Hayward recalls once more:

> 'On the whole, labour relations would have been considered good. Most of the work force had been conscripted from rural areas and had no experience of industry or Trades Union membership. There were many volunteers, too, who were only interested in winning the war. As a whole, the workforce was generally satisfied with the working conditions and wages, but there was a deal of Union activity which many of us found irritating.
>
> 'The chargehand of a section would put the "chit" up on the board to indicate that Nos 9, 12, 21 and 25 – items on the damage report – were done and ready for inspection. On going into the fuselage to inspect the work, one often found that the men responsible for finishing these items had been caught by a shop steward, who was trying to recruit them or persuade them to agitate for higher pay. This inevitably held up the work and made it difficult to deliver the aircraft back on time.

The Stirlingaires: on the extreme left is Charlie Bull, the band's pianist, who had played professionally before the Second World War. The band leader, Vincent Wright, is sixth from the left, standing beside the band's singer, Jean Circuit. (I. Pacey)

'One incident worth recording, although it did not concern me, involved some workers taking waste pieces of Perspex from the scrap bin during a slack period and filing them into brooches or similar ornaments. The foremen turned a blind eye to this activity, it passed the time and did no harm. Suddenly two workers were dismissed for using the time in this way. Others carried on as before. But the two who had been dismissed had earlier refused to join a Trades Union.

'Those of us from rural areas who knew little of Unions could not understand why everyone, including managers, clocked in at 7.15 a.m., while the shop stewards arrived at 9 a.m. and were the only people in the factory not to soil their hands. When Trades Union activity became news after the war, many of us had not formed a favourable impression.'

THE STIRLINGAIRES

Vincent Wright joined the RAF like many other young men at the outbreak of war. But after eighteen months he was posted into industry with about 60,000 airmen up and down the country to make up for the shortage of civilian labour in the aircraft factories. He was fortunate to end up at the SEBRO factory in his home town of Cambridge where he was employed as a fitter (airframes) in the repair of damaged Stirlings. Vincent's talents, however, extended far beyond his regular workaday existence for he had musical talents which added to the enjoyment of his fellow workers and the people of Cambridge. He recalls:

'I had always been a keen musician. My musical career had started back in 1931 when I was asked to play in the Old Dorothy Café Ballroom Band that was then in Sidney Street, Cambridge. My instruments were the saxophone and clarinet. It wasn't long before I had teamed up with a pianist and one other instrumentalist at SEBRO where we used to play regularly in the works canteen for "Workers' Playtime". After a while we added other musicians who were working nearby at Marshall's Airport factory and whom I knew. As a result we named the new band "The Stirlingaires Strict Tempo Dance Band".

'We received no financial support from SEBRO but this did not worry me unduly as we were soon getting engagements in Cambridge and the surrounding district, sometimes five or six times a week. I built the band up on the Glenn Miller theme and this proved very popular. The band at that time consisted of three saxophones, two trumpets, one trombone, a piano, string bass, drums and a lovely lady vocalist named Jean Circuit. I also added a string section of violin, viola and cello for old time music. I gave up the band in 1950 and the bass player took it over but they did not stay together long after that.'

CHAPTER FIVE

Down in Happy Valley

The Ruhr valley was Nazi Germany's industrial heartland from where countless factories supplied a constant stream of weapons and munitions to the armed forces of the Reich. If Bomber Command could either destroy or severely disrupt this vital centre of production, the course of the war might just be altered in favour of the Allies. In what later became known as the Battle of the Ruhr, between 5 March and 24 July 1943 Bomber Command mounted thirty-one major raids against targets in the Ruhr valley – better known to Bomber Command crews as 'Happy Valley'. Deadly belts of flak and searchlights encircled the Ruhr conurbation, making it arguably the most heavily defended area in Germany, apart from the capital Berlin. However, some bomber crews believed their chances of survival were better on the long haul to 'the Big City' than on ops to Happy Valley and its truly awesome defences.

No 3 Group's Stirling squadrons played their part in company with the other heavy bomber squadrons of the Command. But the cost in lives and aircraft was high: Bomber Command lost a total of 981 aircraft and 4,171 crew killed, with a further 1,347 missing. The Battle of the Ruhr was significant, however, in two respects: it marked the opening of the RAF's intensive bombing campaign against Germany, and it showed that the Stirling was beginning to suffer an unacceptably high rate of attrition.

Arthur 'Joe' Edgley, from Dawsmere, Lincolnshire, was a sergeant rear gunner on Stirlings with No 15 Squadron at Mildenhall during May 1943. On the night of 25/26 May the squadron was detailed to bomb Düsseldorf, the first of only two raids flown against this city during the entire Battle of the Ruhr. Joe's crew that night was typical of the cosmopolitan make-up of Bomber Command. They were skippered by Sgt J. Wilson, an Aussie from Sydney; Plt Off B. Cooper, the navigator, was from Concepción, Chile; Sgt P. Arnott, the bomb-aimer, came from Holt, Norfolk; the wireless op, Sgt S. Maxted, from Ilford, Essex; Sgt R. Pittard, the flight engineer, came from Edmonton, London; and the mid-upper gunner, Sgt 'Bud' Seabolt, from British Columbia in Canada. They took off from Mildenhall in Stirling I, BK611:U, 'Te Kooti', at 23.56hrs and headed out across the North Sea.

A groundcrewman perches on 'Te Kooti's' port-inner Hercules engine at Mildenhall in 1943. (D. Mepham)

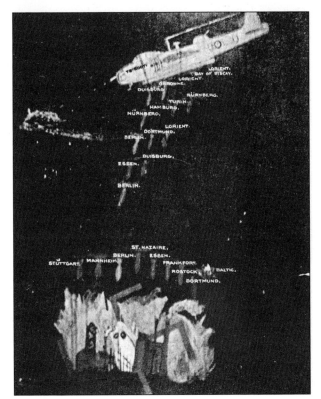

The artwork on 'Te Kooti's' rear fuselage bears witness to its operational history. (D. Mepham)

Returning from Düsseldorf, No 15 Squadron's Stirling I, BK611 'Te Kooti', was shot down and crashed at Grubbenvorst near Venlo in Holland on 26 May 1943. Here are five of the crew, *left to right*: Sgt S.J. Maxted, wireless operator, Sgt P. Arnott, bomb-aimer, Sgt J.O. Wilson RAAF, pilot, Flt Lt B.E. Cooper, navigator, and Sgt Arthur 'Joe' Edgley, rear gunner. Both Arnott and Wilson perished in the crash. (A.W. Edgley)

'After testing my four .303in Browning machine-guns to make sure everything was operating correctly I settled down in my turret and began the long lonely vigil, turning my guns from left to right and searching the dark skies for enemy nightfighters.

'The time was 01.32hrs and we were on the last leg in to Düsseldorf at about 12,300ft, when I saw three or four flak shells burst close to us. Before anything could be done in the way of evasive action the next salvo hit our starboard engines putting them both out of action. The aircraft shook violently and the pilot, Sgt Wilson, told us to prepare to bale out.

'Sgt "Bud" Seabolt, the mid-upper gunner, came on the intercom and reported the starboard-outer engine was on fire with the propeller missing, and the inner engine was minus both the prop and cowling. Bud came on the intercom again moments later asking the pilot for permission to bale out. The skipper said "yes" and out Bud jumped successfully about 5 miles south-west of Düsseldorf.

'Although we had jettisoned the bomb-load the moment we were hit the pilot was having great difficulty holding the aircraft straight and level, so he finally gave the order to bale out. My turret was out of action as it was driven by hydraulic power from a pump in the starboard-inner engine, so I unplugged my oxygen pipe, intercom lead and the electrical connection to my heated flying suit, turned the turret with the hand control to centralize it, then locked it, opened the rear doors and climbed back into the fuselage. I put my parachute on and jettisoned the gunner's rear exit hatch in the starboard side of the aircraft. The slipstream snatched the panel away and I sat down on the side of the aircraft and started to edge myself out. The slipstream hit me with a terrific force and as I poised there, swinging sideways, I could see the aircraft was flying fairly level so, with one mighty effort, I pulled myself back into the aircraft thinking we might be able to carry on and probably ditch somewhere.

'I got back into my turret again and told the pilot that I had been out and struggled back in again. "Well done," he replied, "with a bit of luck we'll make it." Because my turret was out of action I decided to get into the mid-upper turret where I found that everything was working all right. When I rotated it to starboard I could see the damage to both starboard engines and a sorry sight they looked indeed. The two port engines were running quite well although I noticed that the port wing was slightly low and the rudder hard over to port to counteract the tendency to swing to starboard.

'The pilot was showing great skill in keeping the aircraft flying and I asked him what altitude we were at; "9,000ft," he replied. I began to think we might make the coast but we were very soon down to 5,000ft and the skipper told the wireless op, Sgt Maxted, and myself to get rid of as much equipment as possible through the hatches.

'We were still losing height and in no time at all we were down to 3,500ft. "Sorry lads," said the skipper, "I can't keep her airborne any longer, so go to the forward escape hatch and bale out." The navigator, Plt Off Cooper, reported that we had crossed the German-Dutch border so we might now have a chance of evading capture. I went up front and saw the pilot and bomb-aimer, Sgt Arnott, struggling to hold the aircraft. Climbing down the steps to the front escape hatch I found the handle and turned it, but to my dismay it broke off in my hand. I made signs for the others to go back, holding up the broken handle for them to see. As I passed the pilot I could see that we were now down to 1,500ft; I pointed to the rear escape hatch and he raised his hand in acknowledgement as I hurried to the back where I beckoned the navigator to jump. He slid through the already open escape hatch and into the night at 01.46hrs. I bent down to go next but saw Sgt Pittard the flight engineer, so I allowed him to go next instead, which he must have done at something like

When 'Te Kooti' failed to return to Mildenhall on the morning of 26 May, its groundcrew were left wondering as to the aircraft's fate. (D. Mepham)

Luftwaffe personnel sift through the wreckage of a No. 15 Squadron Stirling, BK657, brought down by a German nightfighter over the Dutch town of Portengen, near Utrecht, on 27 April 1943. (Author's Collection)

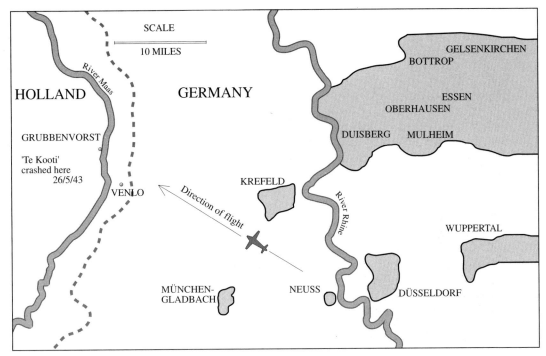

Map to show where 'Te Kooti' crashed at Grubbenvorst on 26 May 1943, in relation to Düsseldorf and the Ruhr valley.

50ft because as I sat down to go there was a terrific rending, tearing crash as the Stirling hit the ground; I covered my head with my arms for protection. The aircraft turned over and over and sideways before finally coming to rest.

'I was still in the rear of the aircraft which was now burning fiercely and I could hear the ammunition exploding all around. Picking myself up I found I could walk all right and shouted to see if anyone else was there. To my delight Sgt Maxted answered, although complaining that his leg was hurt, but after a short time he was able to walk. We removed our parachutes and heavy flying clothing and threw them into the fires before starting to search for the pilot and bomb-aimer. Our search was in vain as wreckage was strewn over a wide area so we decided to make our escape. It was approximately 02.15hrs and we had crashed at a village called Grubbenvorst near Venlo in Holland.'

The narrative that follows, describing Joe Edgley's evasion after the crash, is taken from a transcript of his official postwar debriefing interview (29 August 1945) with IS9 (Western), dated 29 August 1945. IS9 was the top secret section

> No. 15 Squadron,
> Royal Air Force,
> MILDENHALL, Suffolk.
>
> 26th May, 1943.
>
> Dear Mr Edgley,
>
> You will now have received my telegram stating that your son, 1104451 Sergeant Arthur William Edgley, failed to return from an operational flight on the night of 25/26th May, 1943. I am writing to express my deepest sympathy with you in your anxiety, but also to encourage you to hope that he is safe.
>
> He was the Rear Gunner of an aircraft engaged on an important bombing mission over enemy territory, and after take-off nothing further was heard. It appears likely that the aircraft was forced down, and if this is the case, there is some chance that he may be safe and a prisoner of war.
>
> In this event it may be two to three months before any certain information is obtained through the International Red Cross, but I hope the news will soon come through.
>
> Your son had done excellent work in the Squadron and had successfully completed three operational flights. He will be very much missed by his many friends in the Squadron.
>
> His personal effects have been safeguarded and will be dealt with by the Committee of Adjustment as soon as possible, who will write to you in the near future.
>
> May I on behalf of the whole Squadron express to you our most sincere sympathy, and the hope that you will soon receive good news.
>
> Yours sincerely,
>
> Wing Commander.
> (J.D. STEPHENS.)
>
> Mr. C. W. Edgley,
> Gedney Drove End,
> SPALDING, Lincs.

A copy of the letter from the commanding officer of No 15 Squadron to Joe Edgley's father, informing him that his son had failed to return from ops. (A. Edgley)

of MI9 which had run the complex network of escape organizations in France, Belgium and Holland during the Second World War.

'Both Maxie and I decided to evade capture and with the aid of our evasion equipment we began to walk west until about 04.00hrs when we reached a farmhouse, about 5 miles from Venlo and a few hundred yards over the Venlo-Helmond railway line. We approached together and knocked at the windows, one of which was opened by a man. In sign language we did our best to tell him who we were, and he brought us some bread and cheese, but indicated that we were not to enter the house, but to go away.

'After leaving the farmhouse we continued to walk west until 06.30hrs when we arrived at a farm where we took some sacks from the yard to make a bed in a nearby rye field. I found it impossible to sleep owing to being wet and cold and with my mind full of the ordeal we had just been through. We stayed there

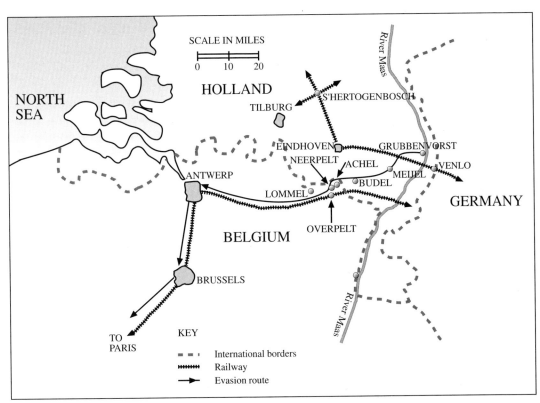

Map to show the evasion route followed by Edgley and Maxted in Holland, Belgium and France between 26 May and 9 July 1943.

until about 07.15hrs. We noticed a small village in the distance and decided to try and obtain help and food. We approached a small house in the village and were given some coffee, bread and bacon. After thanking the people we left to make contact with two men who were working in a nearby field. One of the men took us to his house in the village where we stayed until 13.30hrs on 26 May. When we left the man gave us ample food and drink to last us a few days.

'The reason for us leaving the house was that a person brought a note which was written in English: "Sorry we cannot help you – there are fifteen Germans searching the area for pilots – try to make your way to Belgium – God bless you." The note was accompanied by two pints of coffee, eighteen hard-boiled eggs, 2lbs of butter, cheese, bacon, two loaves of bread and a knife.

'On leaving the house we walked about a mile from the village along a cart track and hid in a ditch where we stayed until 20.30hrs. We then resumed walking along the cart track and met an elderly Dutchman with a bicycle who told us to go to the next farmhouse. Here we were given food, drink and cigarettes and also directions to the Dutch-Belgian frontier. We walked until we entered Meijel at approximately 06.20hrs on 27 May. We went to one house where the woman seemed very scared and who told us to go away. We walked through the village until we saw an old woman whom we approached, but her son said they could not help us. We continued to walk until we came to a farmhouse on the edge of some moors where we remained until 20.30hrs.

'The man whom we had met in Meijel arrived at this farm about 11.00hrs and talked to us. He was a schoolmaster who spoke English and he had brought some civilian clothes. Our outfits were completed by the people at the farmhouse. We wore the civilian clothes over our uniforms and started walking across the moor. After a while the schoolmaster who had given us civilian clothes overtook us on his bicycle. He gave us a map and directions for reaching the Belgian frontier.

'We walked all night and hid in some straw on a farm on the morning of 28 May; we covered ourselves and fell asleep. We awoke to see a farmer milking his cows in a pasture. We waited until he came into the farmyard and then we approached him, giving him quite a start. We explained who we were and he gave us coffee and food, then later that day he and his son took us to another farm. We stayed in a shed all day and that night continued walking towards the frontier. We had with us some food given to us by the farmer. We then met some Belgian corn smugglers who took us to about half a mile north of Budel, close to the frontier. We crossed the Weert-Eindhoven railway line at which point we parted company with the smugglers. Then we found ourselves a hay rick and were soon asleep.

'We awoke at 05.30hrs to find ourselves very wet with the morning dew.

We tried to clean ourselves up but found it impossible and sat down again. We could see the church spire at Budel so we decided to keep it to the left of us, keeping well to the outskirts of the town in the hope that not many people would see us. Just before reaching the frontier we were asked for our passports by a policeman. I thought our time was up. He could understand us and let us continue. At the frontier we met one of the guards dressed in a black uniform and asked him for help. He got us across by going ahead on his bicycle and signalling that the way was clear.

'We went to a farmhouse just north of Achel. We were given food and a man who spoke English was brought to the house. He instructed us to follow another man and we did so until we arrived on the outskirts of Neerpelt. The man, who spoke English, arrived by bicycle and guided us through Neerpelt to the outskirts of Overpelt. After informing us that we were on the main road to Antwerp he left us.

'We walked to a farm just south of Lommel where we spent the night. On the morning of 30 May the farmer bought us railway tickets to Antwerp. We travelled by train to Antwerp where we arrived about 09.30hrs. After we left the station we tried unsuccessfully to find a bridge over the river. About 14.00hrs we went into a park where we approached a man and explained who we were. He took us to an inn where we were introduced to the innkeeper who spoke English. Here we smoked and drank beer. Another man then took us to various parts of Antwerp in an endeavour to find accommodation for us. Before we left the inn we were provided with new civilian clothes by Mrs Cleaver, 13 Falcon Plain, Antwerp, a Belgian woman married to an Englishman, who kept a second-hand clothes shop.

'After being escorted round Antwerp we were taken back to the inn where we were given food, cigarettes, etc. Sometime later Jim Cornelius of Chasse de Merxen, Delene, Antwerp, arrived and took us to his home. He worked in a laundry so he took our clothes to get a good going over. We remained with Jim and his wife until about 8 June. During this period we were interviewed by a man who requested us to fill in forms which would enable him to check our identities with London.

'On the evening of 8 June an elderly woman and a youth came from Brussels and took us back to Brussels with them on the electric train. On arrival in Brussels the youth took us to a man who was standing outside the station and handed us over to him. This man, whom I believe was a Salvation Army captain, took us to M. Calomme Rosset (a Swiss national) of Uccle, Brussels, where we arrived at 21.00hrs. We had supper and during the meal his wife, who professed to be an Irishwoman, questioned us. Apparently she was satisfied with our replies and I stayed at the house until 23 June. Sgt Maxted remained there until 22 June when he was taken to another house. We both met again next morning after I had been taken to the Salvation Army captain's home. We both stayed at this house for about two hours after having breakfast.

'We then left the Salvation Army captain who took us to a small but very clean flat in the middle of Brussels. The flat was occupied by a young nurse or midwife, whose first name was Louise. She moved out to allow us to stay until 14.00hrs on 26 June. During our stay all meals were brought in by the Salvation Army captain and the nurse.

'At 14.00hrs on 26 June two ladies came and took us on a train to the house of Caroline Maes, 24 rue des Hellenes, Brussels, where we stayed until 1 July. On the morning of 1 July we were taken by a lady and a stout man in a car to a place known as the "Captain's house" in Brussels, where we stayed until 8 July. More RAF aircrews came in all the time. During our stay we met the following: Sgt Bill Cole RAF (flight engineer, No 7 Squadron), Sgt Larry Moloney RCAF (rear gunner, Wellingtons), Sgt Goodenough, Sgt Frank Hugo RAF (bomb-aimer, No 7 Squadron, same crew as Cole), Sgt Jack Smith RAF (wireless operator, No 218 Squadron), and Sgt Craven RAAF. [author's parentheses]

'The "Captain" insisted upon us removing all our personal belongings and on 8 July he supplied us with identity cards, passports, railway tickets etc. We were split up into parties. In my party were Sgts Cole, Hugo, Moloney, Smith and Maxted. We were taken to the railway station *en route* for Paris. The journey was uneventful except when we approached the Belgian-French frontier near Mons. By this time we had been joined by a Belgian nurse and a Belgian civilian who was our guide. On approaching the frontier the train was stopped and the passengers were compelled to alight in order to pass through the frontier guards. Our compartment was entered by a German soldier and our guide engaged him in conversation. After a short while the soldier left. We were the only people who did not leave the train. It seemed very funny that they didn't search us.

'The rest of the journey passed without incident. On arrival in Paris at 14.30hrs we proceeded, with our guide, to the underground station, and travelled to what appeared to be the southern sector of Paris where we were taken to a hotel. Sgts Cole and Hugo were taken to another hotel. They rejoined us later, but made no comments. After we had finally settled down in the hotel we were taken out and given a meal by our guides as meals were not provided at the hotel. We returned to the hotel for the night.

'At 07.00hrs on 9 July I arose from my bed in the hotel, feeling quite confident that before many more days had passed I would be in England again. The guide came for us at 08.00hrs and said he was taking us to the railway station *en route* for Bordeaux and Spain. He escorted us for about half-an-hour, when he handed us over to another guide who came up to him.

Informing us that this man was to be our escort, with a few words of thanks the former guide departed. Our new guide shook hands with all of us and wished us "Good morning" and we stayed with him for about 10 minutes.

'In front of us was a long straight road with trees at either side, a wide pavement, and with several streets leading off along both sides. After walking about 200yds our new guide told us to keep close together. As soon as we closed ranks about eight men in civilian clothes armed with revolvers sprung out of one of the side streets and surrounded us. One of them said "Hands up! You are British airmen." Our guide ran off and we did not see him again. We were searched, handcuffed and then marched onto a bus parked down a side road where we found a German soldier at the wheel. We were driven to Fresnes prison near Paris where we were eventually interrogated by the Gestapo.'

Joe Edgley and his compatriots had been captured without any chance of offering resistance. Their escape route had been compromised in both Belgium and France, having been infiltrated by collaborators who betrayed them to the Germans. There were over 1,600 different wartime escape lines in North-West Europe down which evading Allied airmen were passed. The Gestapo tried repeatedly, and often successfully, to infiltrate these lines and break them, dealing brutally with the ordinary people caught with Allied airmen in their homes. The leaders in the escape organizations were captured and tortured by the Gestapo.

The man known as the 'Captain', referred to in Edgley's narrative, was an escape line organizer turned traitor. He was an Englishman whose real name was Harold Cole, a philanderer and con man. Originally an agent for MI9, it is believed he was turned by the Abwehr (German counter-intelligence organization) in about 1942 to inform on the escape line activities in France and Belgium. Eventually, at least 150 people of all ages were betrayed by Cole, and other Belgian and French traitors, who had sold themselves to the Gestapo.

Edgley and Maxted were taken to the notorious Fresnes prison south of Paris which also doubled as a Gestapo headquarters. Behind its dreaded walls were housed mainly political prisoners, spies, saboteurs and members of the French Resistance. Hundreds of men and women were tortured here and died horrific deaths at the hands of the Gestapo. Republican 'loyalist' prisoners, captured during the Spanish Civil War more than four years earlier, were still incarcerated at Fresnes where they served in the prison as forced labour for their German captors.

After interrogation by the Gestapo, the two airmen were eventually transported to Dulag Luft at Oberursel near Frankfurt on 23 August. Here they stayed until they were transferred on the 29th to the prisoner-of-war camp Stalag IVb at Muhlberg am Elbe in eastern Germany, 30 miles north-west of Dresden, where they spent the rest of the war until liberated by the advancing Russian forces on 23 April 1945.

CELEBRATION: Flt Sgt A.M. Halkett DFM (*third from right*) and his No 15 Squadron crew, pose for the camera on 11 October 1942 with N3669, H–Harry, (see page 70) to celebrate the aircraft's sixty-second op. (IWM CH2244)

SPECIAL DELIVERY: The 500–pounder bombs on which people stuck their war savings stamps during London's 'Wings for Victory' week were destined to be dropped on Germany or German-occupied territory. This bomb stood in Trafalgar Square and is pictured here on a bomb trolley at Mildenhall, about to be loaded into the bomb-bay of No 15 Squadron's Stirling, S–Sugar. Also in the picture are the bomb-aimer and pilot who delivered this savings stamp bomb, and the squadron's bulldog mascot 'Bill Prune'. Bill Prune was well-known on the airfield where one of his tricks was to push empty oil drums around the concrete paths and perimeter tracks with his feet, the terrible noise from which could be heard all over the drome. (Author's collection)

WINGS FOR VICTORY

Between 1939 and 1945 the enormous cost of financing the war caused the British Government to place great emphasis on national savings as a means of paying for the fight. The ethic of saving for the national good was promoted to the British public with promises of victory tomorrow in return for a little hardship today. The slogan 'Lend to Defend the Right to be Free' explained the idea clearly and succinctly to the citizens of the 'island fortress'.

As part of a four-month national savings drive between March and July 1943, many cities, towns and villages across the UK staged 'Wings for Victory' appeals in bids to raise money to buy aircraft for the RAF. Organized by the National Savings Committee with the cooperation of the Air Ministry and RAF, much-needed money for buying new aircraft was raised in a concerted nationwide effort by individual homes, streets, factories, offices and schools.

The 'Wings for Victory' weeks were launched on 6 March with the Greater London week which succeeded in raising £162 million. Not to be outdone by their bigger metropolitan cousins, even the smallest country villages held their own 'Wings Weeks'. The tiny Wiltshire village of Bulkington near Devizes, with a population of some 150 people, beat its suggested target of £100 by raising £1,773 6s 5d – an average of nearly £12 a head. By the end of the four-month campaign on 3 July a grand total of £615,945,000 had been raised in England, Scotland, Wales and Northern Ireland. This represented a £70.3 million increase on the similar 'Warships Week' campaign of 1942.

The city of Birmingham's 'Wings for Victory' week between 26 June and 3 July 1943 provided a fitting climax to the campaign by raising a staggering £16,209,630. For 'Brummies', the aircraft manufacturing industry was a big employer and played a large part in the wartime life of this Midlands city. In five years of war between 1940 and 1945, some 620 Stirlings were built in the Austin Motors shadow factory at Longbridge and test-flown from nearby Elmdon before delivery to RAF Maintenance Units up and down the country.

The Short Stirling was a potent symbol of the city's manufacturing might and also of the RAF's strategic bomber offensive against Germany. It was fitting, therefore, that a striking 100ft-tall mural of a Stirling was specially commissioned by the authorities from Joe Woodhall, a graphic artist, to mark Birmingham's 'Wings for Victory' week in the summer of 1943. It Comprised ninety individual panels of painted canvas attached to a scaffolding framework and was erected in Broad Street as the imposing centrepiece of the appeal. Air Marshal Sir Trafford Leigh-Mallory, C-in-C of RAF Fighter Command, opened the event on Saturday 26 June and took the salute before a 5,000-strong marchpast of servicemen representing all branches of the armed forces.

BELFAST: No 15 Squadron's Stirling I, N6065, takes part in the province's 'Wings for Victory' proceedings. (Shorts Plc Neg No: ST539)

MURAL ARTIST: Joe Woodhall (*left*), creator of the Birmingham 'Wings for Victory' mural, is pictured beside a Stirling at Elmdon airport in 1943. (C. Cooper)

H-HARRY: A veteran of sixty-seven raids, Stirling I, N3669, served with Nos 7 and 15 Squadrons and is seen here on display outside St Paul's Cathedral in London during February 1943 as part of the capital's contribution to 'Wings' week. (Author's collection)

'A KEEN SENSE OF HUMOUR', as portrayed by the cartoonist Rowland Emett in the popular magazine *Punch* in 1943. (Reproduced by permission of *Punch*)

In a speech on Thursday 1 July, the high cost of financing the air war against Germany was outlined by Mr Geoffrey Lloyd, Chairman of the Oil Control Board and Petroleum Secretary. Speaking to an audience which included representatives of the Allied governments, he revealed that a Lancaster bomber required about 2,000 gallons of petrol on a raid to the Ruhr and back. A single raid to the Ruhr during this period (involving on average some 600 aircraft) would have used about £200,000 worth of fuel.

While Birmingham held its week-long fund-raising event, RAF Bomber Command was locked into the costly Battle of the Ruhr. During the period of 'Wings Week', from 26 June to 3 July inclusive, Bomber Command mounted three major raids against targets in the Ruhr valley as well as a number of minor operations, losing ninety-one aircraft and crews in the process, eighteen of which were Stirlings. The cost of replacing these lost aircraft and their highly trained crews quickly wiped out the £16 million raised in Birmingham. The cost of the war was indeed very high in every respect.

LIFE-SIZE: The city of Birmingham's impressive Stirling mural was displayed in Broad Street during the week of 26 June–3 July 1943. (C. Cooper)

WINGS FOR VICTORY

CHAPTER SIX

Life at Ground Level

A close bond of trust and mutual respect existed between air- and groundcrews, and on Stirling squadrons it was certainly no different to any other Command of the RAF. After all, the safety of an aircrew once they boarded their aircraft depended upon the skills of their groundcrew. From the moment a pilot started up to the minute he landed his aircraft back at the airfield, it was reassuring for him to know that the servicing of his aircraft was in the hands of a dedicated and meticulous team of specialist tradesmen. As one pilot commented, when you are facing flak and nightfighters over Germany, the last thing you need to worry about is whether or not your aircraft has been serviced properly.

Following an engine change, for example, a Stirling would be taken up for an air-test. The pilot and flight engineer would frequently ask their groundcrew engine fitter to grab his toolbox and come along – unofficially of course – and a spare parachute was always found for him. An ex-Halton engine fitter had much more engineering experience than a flight engineer straight out of training at St Athan, so it usually made sense to listen to his wise counsel.

LAC Len Marsh was a fitter (aero engines), the most skilled groundcrew trade, based with No 7 Squadron at Newmarket and Oakington during the spring of 1941. Below he recalls some of the teething problems experienced with the Bristol Hercules II and XI engines fitted to the earlier Stirlings:

'In the early days we had plenty of snags with the engines, some of which were dangerous. An example of this involved the dry sump which all Stirlings were fitted with. It was a 25-gallon oil tank fitted at the rear of each Bristol Hercules radial engine and situated between the bottom two cylinders. This oil was continuously filtered and pumped around the engine. As soon as the sump was full with the excess oil circulating around the engine it was pumped back to the main tanks. This process enabled the Stirling to engage in violent fighter-like manoeuvres without suffering oil starvation to the engines.

'When a Hercules engine was shut down the sump may have been only half full. Once the remaining oil had drained down it would have filled the sump. Any surplus oil drained into the bottom two cylinders and there it would remain until the engine was started again and run-up. Then the piston tops would hit this oil, the snifter valves (pressure valves) fitted to each of the cylinders would open and allow the oil to escape, thus preventing any damage to the engine. What the designers seemed to have forgotten was that when the viscosity of the oil thickened during the first spell of cold weather, the snifter valves could not cope with it and the pistons, upon hitting this solid mass of oil, kept on going. The result was one terrific bang as the two piston heads came through the cowlings to bury themselves in the ground under the engine.

'It was not long before a modification was introduced. A long handle was made so that the fitter could turn the engine over by hand which gave the snifter valves sufficient time to operate. The task took one hour to perform but after that we never had any more trouble.

'The Bristol Hercules engine was what was known as a sleeve-valve radial. It needed very little maintaining and, being air-cooled, its exhaust shroud was circular and in front of the engine. It looked like a large ring. When the pilot was waiting for his take-off signal, with the throttles right forward for maximum power, these four rings would start to glow from red to yellow and could be seen for miles. It would get even worse at high altitude where there was less air to cool them down – a perfect visual guide for an enemy nightfighter. We were constantly painting these rings with a special paint which was supposed to prevent the glow. However, it was always flaking off so it is difficult to say if it really made any difference.'

The Stirling's great height above the ground presented a whole set of new problems when it came to servicing:

'In the beginning we had very little in the way of trestles to get up to the engines for servicing. In order to undo the cowling buttons on the engines one would have to sit astride the top. What trestles we did have were originally used for twin-engined Wellingtons and were next to useless in trying to reach up the 20 or so feet from the ground to the Stirling. When the new engine gantries arrived they were found to be pretty useless, too. About 20ft high, they were made of tubular steel with one tube sliding within the other, allowing adjustments to be made to the height. One needed the agility of a monkey to reach the top where there was a wooden platform measuring about 3ft by 6ft. This did not give very much room for a cowling and a tool box; with oil spillages and rain it soon became like a skating rink. However, it did have a hand rail which was the only good thing about it.

'To reach the top of the structure a metal ladder was attached below the edge of the platform, but it was rarely at the correct angle and to adjust it to

Fitters at work on a Bristol Hercules engine at Tilstock on 27 March 1944. (IWM CH12932)

the height required was a work of art in itself. With the weight of the platform on top pressing down and the length of the ladder itself making accurate adjustment virtually impossible, it was very difficult to get it even. The whole structure was carried on two large car-type wheels. When you were working at the top, the platform would begin to sway and the wheels would then decide to move. It was a standing joke at Newmarket that our needs for a parachute were greater than those of the aircrew.'

In order to keep the aircraft fully serviceable and to enable the squadron to mount a maximum effort raid if called upon to do so, there were a number of other crucial tasks that had to be performed by the groundcrews. Some were messy and others uncomfortable, while a few were looked upon as being downright dangerous:

'An engine could often develop an oil leak and, assisted by the slipstream, the oil would spread over a very large area behind the engine. If during a raid the aircraft was caught in a searchlight beam it would shine like a mirror and this was not appreciated by the aircrew. It was a most unpleasant job having to wash the oil from under the wing, which meant lying on your back on a trestle with a bucket-full of petrol and plenty of rags. You can imagine how wet it became with the petrol running up your arms while the 100 Octane petrol itself would leave a white deposit on your skin and sting like fury. The aircrew would sometimes help by filling your buckets for you.

'During the winter when ice and snow was around, the unpleasant task of de-icing the wings would have to be carried out. This involved sitting on the wing with a rope tied around one's waist, wearing waterproof overalls (which did not remain waterproof for long). A de-icing fluid was sprayed on to the wing which was only effective for about four hours. Sometimes the aircraft would need to be sprayed twice before take-off. The leading edge was covered with a brown putty-like substance called "Kilfrost" de-icing paste. Over a period of time it gradually became thin and it seemed that all one's spare time was needed to replace this paste. I once heard of a groundcrew member whose waist rope was cut by one of the barrage balloon cutters fitted in the leading edges of the wings. He promptly slid over the edge of the wing and on to the ground. Luckily for him he was near the wing root at the time so did not have far to fall.

'The Stirling had seven petrol tanks in each wing, some of which were near the trailing edge. To fill them one had to lie on the wing because it was too steep an angle to stand up. During the winter with the frost and ice one tied a rope around one's waist and anchored the other end to the undercarriage. This helped to stop you from sliding down over the edge.

'One did not always have to fill a tank right up since the amount required was measured by a gauge and a dipstick. However, a gauge could sometimes develop a fault but a dipstick never did. The individual amount of fuel was recorded on the aircraft's Form 700 – the aircraft's "Bible". Everything that was done to an aircraft was recorded and signed for in this book. The Stirling's fuel system was very good and all four engines could be run from any tank in either wing. In the event of a tank being holed by enemy action the remaining petrol in the tank could be pumped into the others. The horizontal level would have been lost but this could be corrected by moving petrol from one tank to another.

'Lack of proper maintenance equipment often caused a lot of trouble with servicing. The oleo legs on the undercarriage needed daily attention and consisted of two metal sleeves, one contained within the other, filled with a mixture of very thin oil and air. Under pressure this caused a spring-like action which helped the undercarriage to soak up the bumps on landing. The

LAC Len Marsh, groundcrew engine fitter with No 7 Squadron. (Mrs J. Marsh)

A Matador petrol bowser is used to replenish the fuel tanks of a Stirling I. (*Aviation Photo News*/Brian Stainer)

LIFE AT GROUND LEVEL

Stirling Is of No 7 Squadron are refuelled and bombed-up in preparation for ops during 1942. (IWM CH5282)

rocking of the aircraft during the course of a day or night caused the legs to creep and sink so they needed to be topped up each day by the airframe fitter to bring them back to their normal height.

'Two cylinders were left on the ground at the edge of the dispersal pan. One was filled with compressed air, the other with oxygen. The air was used for topping up the oleo legs on the undercarriage, the oxygen for replenishing the small mobile oxygen bottles at various positions in the aircraft, used by the crew when moving around inside during flight. Having no trolleys to move them they were pushed across the ground by your feet and before long they were covered with mud. In such a condition it was difficult to see what the colour-code bands were (compressed air – yellow; oxygen – white/blue) which identified the contents of the cylinder. The danger arose when, instead of compressed air, an oxygen cylinder was connected to an oleo leg to top it up: oil and oxygen together create a dangerous explosive mixture. To my knowledge this happened at least once on No 7 Squadron, but luckily it was caught just in time for the fitter to turn off the cylinder valve and drain off the offending gas. He was a very worried man but luckily for him – and the Stirling in his charge – all went well.'

THE ACHILLES HEEL

From the moment of the very first manned powered flight in 1903, arguably every aircraft that has flown since then – whether combat or civil – has an Achilles heel. In the case of the Stirling this design weakness was inadvertently added when its undercarriage was lengthened to increase the angle of attack to counteract its excessively long take-off runs. The tall gangling undercarriage quickly became the Stirling's trademark, but its vulnerability to collapse under strong side-loadings was only half the story, as Sgt J.H. Spiby, an RAF aircraft acceptance test pilot at Stradishall and Chedburgh during 1942, relates:

'The considerable height above ground, and the rake, of the pilot's seat made it very difficult for the pilot to accurately judge touchdown attitude and precise ground position in the "stalled-on" three-point attitude (as per the book). So it was quite common for a stalled aircraft to be "dropped" a foot or two on landing. This shock loading (almost imperceptible) invariably twisted the mainplane structure, rendering the aircraft a rogue to fly thereafter since the safe take-off, climbing and stall speeds would be increased by some 25–35mph and catch out even the best and most experienced pilots.

'On landing there was virtually no rudder control below the three-point touchdown speed (75–85mph) so in the event of a swing, due to, say, deceleration, cross-wind, uneven tyre pressure or single-wheel touchdown, in order to stay on the runway engine power had to be used if the runway was long enough, or hard differential braking otherwise. Since the brakes were generally unreliable at speed it was very easy to run out of options and the aircraft was destroyed through no fault of the pilot.

'For a combination of these reasons it became common to adopt a much higher approach and landing speed (130–140mph) in a tail-up attitude and "fly" the aircraft on to the ground at a much higher speed than in the book. This higher-than-designed speed naturally caused the tyres to creep round on their rims and pull out the valves sooner or later. A burst tyre on landing or take-off would create a virtually uncontrollable swing and destructive ground-loop, either on that flight or thereafter.

'As already mentioned the Stirling's wheel brakes were unreliable, but the cause of high-speed brake failures owed more to working practices than to poor design. These high-speed failures were invariably due to the re-use of oil-contaminated brake-linings which had been cleaned and processed as per the

The Stirling's undercarriage was a complicated structure, retracted (or lowered) in two operations by electric motors which were prone to breakdown. It is seen (*above*) fully extended and (*below*) in the second stage of retraction into the nacelle. (Shorts Plc)

One of a pair of weak joints on each undercarriage assembly which could cause the whole stalky structure to collapse under strong side-loadings. (G. Blows)

book, but which would continue thereafter to sweat the contamination each time a certain cool temperature was exceeded in high-speed use. This would result in either uneven (swing-making) braking or none at all, hence the wholesale destruction of good aircraft and crews. I was utterly unable at any official level to get this practice stopped.'

Another unique feature of the Stirling was its twin castoring retractable tailwheels. In common with other tail-dragging aircraft, tailwheels often set up a violent side-to-side movement (shimmy) on take-off and landing which, apart from causing extreme discomfort to the tail gunner sitting above them, could also cause serious damage to the aircraft structure. This problem was eventually cured by the development at Farnborough of a new tyre which had two heavy tread ridges in contact with the ground, the Aero Tail Twin Contact Tyre. (via Mrs N. Curtis)

A runway overshoot at Rheine in Germany on 18 April 1945 caused this No. 295 Squadron Stirling IV to roll into a bomb crater, damaging its starboard undercarriage. The pilot of N-Nan, serial LK 132, was Flight Lieutenant Ron Sloan. (J. Earnshaw)

Flt Lt Ken Marshall and his No 199 Squadron crew do not appear too perturbed after their Stirling III, P-Peter, burst a main-wheel tyre on landing and swung off the runway at Manston, late in 1944. (K. Marshall)

Landing at Buckeburg, Germany, on 28 December 1945 with a cargo of mail, No 570 Squadron's Stirling IV, PW440, burst its port main-wheel tyre and the undercarriage collapsed. The aircraft was subsequently written off. (Military Aircraft Photographs [MAP])

CHAPTER SEVEN

'A Piece of Cake!'

The ever-present threat of sudden injury or death at the hands of German nightfighters was very real to the crews of Bomber Command. Riding the night skies with a deadly cocktail of high explosive bombs and full petrol tanks, bombers were potentially as lethal to their crews as to the intended recipients of their bomb-loads.

Whenever the RAF flew night operations against targets in Germany and occupied Europe, Luftwaffe nightfighters were scrambled to stalk and intercept the RAF bomber stream from which they took a heavy toll in aircraft and lives. Pumping the bombers' bellies full of cannon shells and then raking them with batteries of machine-guns, the nightfighter crews watched as the hapless leviathans exploded in balls of incandescence and fell away to earth.

One of the many young Luftwaffe airmen defending the Reich against the British *Terrorfliegers* was Wilhelm Johnen, a Messerschmitt Bf 110 nightfighter pilot with 1/NJG 1. His account of a typical nocturnal encounter with two RAF bombers in the summer of 1943 puts the other side of the story:

'Shortly after one o'clock a four-engined Halifax crossed my path. I attacked immediately with no heed for the defence and fired at its petrol tanks. The bomber exploded and fell to earth in a host of burning fragments. 01.03. Five minutes later I saw a pair of huge shark fins just below me. I had already recognized it as my old friend the Stirling . . . The enemy bomber grew larger in my sights and the rear gunner was sprayed by my guns and silenced just as he opened fire. The rest was merely a matter of seconds. At 01.08 this heavy bomber fell like a stone out of the sky and exploded on the ground.'[1]

Between July 1942 and May 1945 it is estimated that the RAF lost 2,278 bombers on night operations to German nightfighter attacks, with a further 1,728 damaged in shoot-outs, together representing the biggest single cause of RAF bomber attrition for the entire war. Overall, a total of 7,449 Bomber Command aircraft went missing on night operations between 1939 and 1945.[2]

The night of 12/13 August 1943 proved significant for several Stirling crews: one became the victim of 'friendly' fire, while another encountered a roving

Messerschmitt Bf 110 and only just escaped destruction. On this night a mixed force of 152 Stirlings, Halifaxes and Lancasters of Nos 3 and 8 Groups crossed the Alps to bomb the city of Turin in northern Italy. Two aircraft failed to return, both of them Stirlings. One was a No 218 Squadron machine which was shot at and badly damaged by a 'friendly' bomber while crossing the Alps. Although mortally wounded in the attack its pilot, Sgt Arthur Aaron, flew the crippled bomber south over the Mediterranean Sea to crash-land in Algeria, an act of tremendous endurance achieved against all the odds for which he was awarded a posthumous VC.

Another Stirling, which almost became the third to FTR that night, was a No 149 Squadron aircraft skippered by Plt Off Jim 'Red' Gill. He and his crew had joined the squadron in July via No 11 OTU, Westcott, and No 1651 HCU at Waterbeach. This trip to Turin was to be their thirteenth operation – and an unlucky thirteenth it certainly proved to be. They left Lakenheath's runway at 21.14hrs flying in Stirling III EH904:K-Kitty but they were attacked by a German Messerschmitt Bf 110 nightfighter north of Paris on the

Jim 'Red' Gill's crew, pictured at No 1651 HCU, Waterbeach, in May 1943. *Back row, left to right*: Les Beaton, bomb-aimer, 'Curly' Mason, rear gunner. *Front row*: Harry Cosgrove, navigator, Jack Atkins, wireless operator, Jim Gill, pilot. (J. Atkins)

outward leg at about 22.37hrs. Sgt Jack Atkins, the WOp/AG, takes up the story:

'Briefing was early; we all knew the tanks were full and the bomb-load scanty – obviously a long trip in store. Berlin had been the order of the day for the past couple of weeks so when the magic curtain was pulled and we could see the red ribbon leading to Turin there were smiles of relief all round. Sgt "Curly" Mason, our rear gunner, who always had something to say, loudly remarked "a piece of cake!". The icing was skimmed off slightly when the CO informed us we were being routed through the Alps and not over them like the Lancs and Halibags. As I remember, some of the top peaks reached the 15,000ft mark or more, and most Stirlings with a full load had no chance of reaching that giddy height. Poor old K-Kitty never even topped the 12,000ft mark! This, however, was a minor irritation compared to a Berlin outing and the German nightfighters didn't put themselves out defending Italian targets.

'With letter-writing and eggs and bacon behind us, it was into the pick-up wagons and out to the respective dispersal points. Here we met the friendly local villagers who, unbelievably, were allowed to get within a few yards of the bombed-up planes. There was very little in the way of security: a main road cut across one corner of the drome and all crews had their village friends bid them, "Have a good trip." I really do believe on some occasions they even knew where we were going!

'Take-off and setting course all went according to plan, with two squadrons of something like forty planes leaving Lakenheath *en route* for Italy. Over the sea all gunners tested their guns and it was still light enough to see other Stirlings outward bound on the same track. I had tuned in my R1156 receiver to base in case of a recall which seemed very improbable – weather and K-Kitty were both on best behaviour.

'By now we were well into France and somewhere to the north of Paris. This area was not a usual hunting ground for nightfighters but there was one out tonight and he managed to find us.

'I remember a noise like an express train going through a station as tracer and cannon shells were pumped into the rear of Kitty from slightly below and out through the front. Sparks were everywhere: the bundles of Window I had beside me were smouldering from tracer that had passed between my legs and more tracer had set light to the oxygen bottles. By this point the flames were beginning to get out of hand. "Cossy" [Sgt H.J. Cosgrove], the navigator, grabbed a fire extinguisher but pulled the trigger too quickly and white foam sprayed everywhere, covering the bomb-aimer, Les [Sgt L.J. Beaton].

Stirling I, N6122:Q, 'East India 2', joined No 149 Squadron in December 1941 but was written off in a landing accident in June of the following year. (Author's collection)

No 149 Squadron's N-Nuts, a somewhat weather-beaten Mk I, in flight. (Author's collection)

'Jim "Red" Gill, our New Zealander skipper, had pushed the nose down hard. He told us afterwards that he'd given the order to bale out but we didn't need any telling. I clipped on my chute, always kept handy on my desk, but I could hardly get out of my seat due to the steepness of the dive. I just managed to get to the front exit where I saw Les jumping up and down on the hatch which had been jammed shut by a cannon shell. Butch, the flight engineer [Sgt H. Wickson], was next in line, then Cossy and finally me – number four in a queue that wasn't moving and didn't seem to have a future – so I crawled to the rear escape hatch.

'Butch had decided he could help Jim get Kitty on a level flight and between them they slowly pulled us out of the dive. Cossy said later we were down from 11,000ft to something around 700ft. He also confided to me that he had clipped his chute on back to front – the handle firmly against his chest. A good job Les couldn't get the hatch open.

'"Smokey" [Sgt M. Levett], our mid-upper, had switched on the fuselage lights and what with that and the fire we were lit up like a Christmas tree. He was sitting with his feet dangling through the escape hatch into the slipstream as he gave me the sign that he was going to jump. I grabbed his shoulders as the plane was now flying straight and level and we had some chance of getting back home.

'Up front things were improving. Les had given up all ideas of jumping, Cossy had the fires under control and Jim was fairly happy we were maintaining our height. The bomb-load had been jettisoned near Chartres at 23.30hrs and now, flying at 180 degrees to our outward course, we tried to sort ourselves out. Jim had a quick roll-call and all checked in except Curly. Smokey was back in his turret so I made my way back to the rear. The last couple of yards got really tight. To me, all rear gunners were heroes: how they got in was an effort; how they got out again was a miracle.

'There was no way I could get the turret doors open and there were jagged edges everywhere. Curly was slumped over his guns, the front of his turret was smashed completely and it was freezing cold. I tried to get my arms around him but it just wasn't possible. It looked as though he was in danger of falling out.

'It was still a long way from home and we had a lot of sea to cross when Butch reported the fuel gauges were US [unserviceable]. Jim told me to get an SOS off and for Cossy to give the tracking station a course to track us home. We crossed the French coast just east of Cherbourg and headed due north. I'd hardly got the first SOS off when I had a reply from somewhere in England. In the shortest of time I sent Cossy's course and it was acknowledged: from then on we were being tracked in and we all felt a lot better. Speaking to other WOps who'd found themselves in trouble, all had nothing but praise for those civilian wireless operators who helped them over the last few miles to the English coast. By now we were heading directly towards Portsmouth – not safe but safer – and we could hear the Portsmouth balloon barrage signals in our headphones. Jim, not knowing how much damage Kitty had taken, decided against any quick manoeuvres and so we kept on going. Petrol appeared to be holding out so we made for base and with nothing in the way of formalities we went straight in to land. The undercarriage held up (a rare event on a shot-up Stirling) and we rolled to a stop. The time was 01.49hrs.

'We were all out of the aircraft in seconds and ran round to the back where we lifted Curly out of his turret. He was dead and had been almost cut in two by the cannon fire from the nightfighter. In a way it was a small consolation to us knowing that he could have felt and known nothing.

'Poor old Kitty was soon surrounded by curious onlookers of all ranks and

Returning from a mining operation in the Heligoland/Frisian Islands area, Sgt J.A. Jerman's No 149 Squadron Stirling I, N3752:O, hit a hill near Åbenrå on the Danish mainland, some 30 miles north of Flensburg, on 18 May 1942. The crew became prisoners-of-war. After laying its mines from a height of 600ft another squadron aircraft, N6080, piloted by WO Ashbaugh, dived down to 200ft where it shot up five German E-boats before turning for home after sustaining damage to its starboard-outer engine. (via Chaz Bowyer)

Captained by Squadron Leader G.A. Watt, No 149 Squadron's Stirling I, BF325:A, was detailed to attack the German city of Mainz with a load of incendiaries on the night of 12/13 August 1942. Watt and his crew were unable to pinpoint the target so they jettisoned their bomb-load over the Eifel mountains west of the German town of Cochem. Shortly afterwards they were attacked and damaged by a Messerschmitt Bf 110 nightfighter over the Ardennes forest in eastern Belgium. Struggling homewards, damaged and short of fuel, they made the coast of England, but when their engines cut out they were compelled to make a forced landing perilously close to some houses in Rumfields Road, Broadstairs, Kent, early on 13 August 1942. (via D. Collyer)

the next morning numerous photographs were taken of her from every angle. Soon, copies bearing two captions were sent to all Stirling squadrons to be posted on the Mess notice boards: one above stressed the capacity of the Stirling to absorb great damage and still return to base, and underneath in large print the legend "DON'T STRAGGLE – THIS COULD HAPPEN TO YOU". We didn't ever straggle, we just couldn't get as high or as fast as the others, but that was not our fault. K-Kitty, God bless her, was just made that way.

'We were given (very reluctantly) a day off. I had a rollicking from the station wireless officer because I had forgotten to wind in my trailing aerial, for which I was fined one shilling. As we lifted him on to a stretcher I took an empty orange juice tin from Curly's pocket: it was full of shrapnel holes. I have it to this day. A label on the bottom says, "CURLY 13th AUG 43 – A PIECE OF CAKE".'

THAT OLD-FASHIONED STIRLING OF MINE

This affectionate song was often sung in the Sergeants Mess at Lakenheath during 1943 when ops were off and the beer was flowing freely. Contributed here by Jack Atkins, a sergeant WOp/AG with No 149 Squadron at Lakenheath in the spring and summer of 1943, it is sung to the tune of *That Old-Fashioned Mother of Mine*.

> Just an old-fashioned Stirling
> With old-fashioned ways
> A fuselage tattered and torn.
> Four Hercules engines keep chugging away
> She's flying from midnight to dawn.
> Though she don't go so fast,
> No great height does she claim,
> Sure there's something that makes her divine:
> When she flies there on high
> She's the Queen of the sky,
> She's that old-fashioned Stirling of mine.

(Two more songs parodying the original tune and words were once in the RAF's unofficial musical repertoire: *That Old-Fashioned Avro of Mine* from the interwar period referred to the Avro 504 biplane trainer, and *Old-Fashioned Wimpy* was sung by No 70 Squadron crews from 1941 in the Middle East.)

1. Wilhelm Johnen, *Duel Under the Stars*, p. 105 (Crecy Books Ltd, 1994)
2. Sir Charles Webster and Noble Frankland, *The Strategic Air Offensive Against Germany*, Vol IV, pp. 437–9 (HMSO, 1961)

CHAPTER EIGHT

For Gallantry

In the closing months of the Crimean War in 1855, the Conspicuous Gallantry Medal (CGM) was first instituted for NCOs of the Royal Navy and Royal Marines as an award for 'exceptional gallantry in the face of the enemy'. Second in status to the Victoria Cross, the CGM was the NCOs' equivalent to the Distinguished Service Order (DSO) which was awarded only to officers.

Over eighty years later, during the Second World War, on 10 November 1942, the CGM was extended to Army and RAF personnel 'whilst flying on active operations against the enemy'. Yet, by the war's end, only 110 awards of the CGM to RAF and Commonwealth NCO aircrew had been made compared with a total of 1,087 DSOs to RAF and Commonwealth officer recipients (not including eighty-five bars to the award). Since some 70 per cent of Bomber Command aircrew were NCOs,[1] it is somewhat incomprehensible – in spite of the fact that the CGM was not instituted until halfway through the war – that they should account for less than 10 per cent of the combined total of exceptional gallantry awards made.

Of the eighty-nine awards of the CGM to members of Bomber Command during the Second World War, eight went to Stirling aircrew[2] of which three were won in dramatic circumstances over Berlin on a single night, 23 August 1943. This date marked the opening raid of what later became known as the Battle of Berlin in which Bomber Command despatched nineteen major raids to the German capital between August 1943 and March 1944. In more than 10,000 aircraft sorties over 30,000 tons of bombs were dropped on 'the Big City' for the loss of 607 aircraft and 3,347 men killed.

Flt Sgt Gil Marsh and his No 622 Squadron crew flew on this first raid of the battle and came very close to becoming an early fatality statistic in the casualty returns. The outstanding heroism of Marsh's bomb-aimer, Sgt Jack Bailey RCAF, earned him the CGM. George Wright, who was the wireless operator in Marsh's crew, relates the dramatic events of that night:

'We did four ops between the beginning of July and 9 August 1943 with C Flight of No 15 Squadron. On 10 August, C Flight became A Flight No 622

Sgt Gil Marsh, pilot of BK816, was seriously wounded over Berlin on 23 August 1943 when a cannon shell from a German Ju 88 nightfighter exploded next to him in the cockpit. (G. Marsh)

Squadron and on that night we went to Nuremberg, then to Turin on 12 and 16 August, followed by Berlin on the 23rd.

'We took off at 20.30hrs in Stirling III BK816:B and the trip took 8hrs 45min. The crew was Flt Sgt Gil Marsh, pilot; Plt Off R. Richards, navigator; Sgt Jack Bailey, bomb-aimer; Sgt George Wright, wireless operator/air gunner; Sgt Jimmy Meaburn, flight engineer; Flt Lt F. Berry, gunnery leader standing in for Sgt W. Smith as mid-upper gunner; and Sgt Art Hynam, rear gunner.

'The journey out was different from our previous trips, with no searchlights at the enemy coast, nor flak or nightfighters. It had been decided to bomb at a lower level than other Stirlings, at about 12–13,000ft. As we approached the target area we could see the Pathfinder marker flares and then suddenly a complete ring of searchlights around the city pointing vertically upright, with some coning aircraft as they passed through this barrier. Whenever I recall the raid this feature always springs to mind. It had so many searchlights that it was as good as daylight even though it was approximately midnight.

Sgts Art Hynam, rear gunner (*left*), and Gil Marsh. From his vantage point in the rear turret, Hynam gave the order to take evasive action after BK816 was attacked by the Ju 88. (G. Marsh)

'We went through the searchlight barrier without being coned but could see other aircraft in difficulty with the coning, and with nightfighters coming in from above. We dropped our bombs on the target but just as our bomb-doors were closing we were attacked by a Ju 88. When the call came from Art Hynam to take evasive action our aircraft was sluggish to respond because the doors had not fully closed by that time.

'Different colours shot through the starboard side into the flight deck. We were attacked three times; the hydraulics of the rear turret were put out of action during the final attack, but not before the enemy aircraft was seen to fall away and it was claimed as a possible. There was damage to the tailplane, the port elevator was shot away and because the port-outer engine was shooting flame it was feathered. During the attack a cannon shell appeared to hit the edge of the armour plate to the pilot's seat and explode. Gil was hit in some six places around the groin and leg, cutting the sciatic nerve.

'The aircraft fell rapidly as Gil slumped over the control column. As a result

The bravery of Marsh's Canadian bomb-aimer, Sgt Jack Bailey (*left*), earned him the immediate award of a Conspicuous Gallantry Medal (CGM). On the right is George Wright, the wireless operator, whose recollections of the fateful Berlin raid form the basis of my account. (G. Marsh)

Jack Bailey, still in the bomb-aimer's position, took a knock and was temporarily stunned. Dickie Richards and I pulled back on the control column either side of Gil until the aircraft levelled out at 1,500–2,000ft. Gil Marsh eventually recovered consciousness sufficiently to get the aircraft back up to about 4,000ft. Dickie Richards then took the co-pilot's seat by which time Jack Bailey had recovered and came up to the flight deck from down below. He took the place of Dickie and I removed Gil from the pilot's seat and propped him behind it on the floor against the side of the fuselage. Jack Bailey then occupied the pilot's seat. Dickie injected Gil with morphine carried in the first-aid kit and used up all that we carried. We made him as comfortable as possible and then he lapsed in and out of consciousness for most of the time, muttering from time to time that he was thirsty, so we gave

Gil Marsh's unassuming log book entry for 23 August records 'War Operation Berlin' while the gunners' report of the incident with the Ju 88, pasted on to the righthand page, reveals the grim reality of that night. (G. Marsh)

BK816's groundcrew, pictured at Mildenhall in the summer of 1943. Sgt Sid Hiscock, engine fitter, i/c groundcrew, is second from the right, Tony Godfrey, airframe fitter, is third from the left. The others are unknown. (G. Marsh)

him all the coffee we could muster from our flasks. All this time we continued to fly at 4,000ft on three engines. I spent time in the astrodome looking for enemy fighters in between listening out for the quarter-hourly broadcasts and looking in at Gil as opportunity allowed.

'As we made headway it was slowly getting lighter and we were concerned that soon we might encounter enemy fighters. The crew had a short discussion on the intercom and agreed that only in dire necessity would we bale out or ditch because of the critical condition of Gil; he was in terrible pain.

'As we crossed the coast of Denmark into the North Sea we were not absolutely certain of our position so we decided to break R/T silence. I tried to get through to base for a fix but the airwaves were very cluttered and noisy. I then sent out an SOS which cleared the lines and then requested a fix for base which I was told later was 420 miles, followed by two further fixes *en route* to get us home.

'We were now in daylight over the North Sea with a fixed route for base so it was decided to try and start the port-outer engine. This was successful amid a great cheer that went up from the crew, except for Gil who was still unconscious. Now we felt that we were on our way home. Another discussion took place as to whether we should try and land at the nearest aerodrome or continue to Mildenhall. It was decided that because Mildenhall was aware of our condition, then we should go on to our home base.

'When we arrived at Mildenhall our fuel was very low and Jack Bailey received instructions from flying control to go straight in to land. Everything was cleared in case our undercarriage malfunctioned. Escape hatches were removed and we took up crash positions. Jack was talked down by flying control and he made a very good landing. Apparently we had only 75 gallons of fuel left in our tanks.

'With the help of the ambulance crew we got Gil out first. Flt Lt Berry suggested that only he, Jack and Dickie should attend the debriefing and the rest of us ought to see Gil off in the ambulance. When we got out of our Stirling there was Sid, our sergeant fitter for BK816, with tears running down his rosy cheeks. He was very emotional at our return some 1½ hrs later than the other Mildenhall aircraft which had all landed. He was so pleased to see us.'

Jack Bailey received an immediate award of the CGM and a telegram from 'Butch' Harris with arrangements to return to Canada for a pilot's course. Gil Marsh remained in hospital until June 1944 and then returned to Mildenhall, but not to fly again as a pilot. By request, the rest of the crew was left together. They obtained a new pilot, Flt Lt Stoddard, and a new bomb-aimer, Flg Off Ashton (another Canadian), and Flg Off Burroughs (NZ) became their new navigator. Dickie Richards was awarded a DFC but flew on only two or three more ops before going sick. George Wright received his commission some six weeks later after an interview with Air Vice-Marshal Don Bennett (AOC PFF) at which he only discussed with him the details of this op on 23 August 1943. Art Hynam received his commission later. The crew then went on to finish their tour on 28 May 1944, although George Wright did fly more ops than any of the rest of the crew, finishing on a total of thirty-one.

1. M. Middlebrook & C. Everitt, *The Bomber Command War Diaries* (Viking Penguin, 1985), p.712
2. Quoted in A. Cooper, *In Action with the Enemy* (William Kimber, 1986): Flt Sgt J. Russell RCAF, pilot, 15 Sqn (03–04/10/43); Flt Sgt O. White RNZAF, pilot, 75 Sqn (23–24/08/43); Sgt O. Jones, F/E, 90 Sqn (22–23/09/43); Flt Sgt E. Durrans, WOp/AG, 90 Sqn (10–11/04/44); Sgt H. Donaldson, WOp/AG, 199 Sqn (05/03/44); Flt Sgt A. Larden RCAF, B/A, 218 Sqn (12–13/08/43); Sgt J. Bailey RCAF, B/A, 622 Sqn (23–24/08/43); Sgt B. Bennett, WOp/AG, 623 Sqn (23–24/08/43).

CHAPTER NINE

The Old Man with the Scythe

At some 13,000ft, the poor altitude performance of the Stirling was a contributory factor to its high attrition rate, but it did give its bomb-aimers a much closer – and altogether more startling – view of the scenes of destruction unfolding below them. This altitude handicap also served to emphasize the Stirling's vulnerability to the flak streaming up from the seas of flame below, and from the prowling German nightfighters above. Danger also lurked in the shape of bombs dropped from the higher flying Lancasters and Halifaxes which rained

Newly arrived on No 214 Squadron in August 1943, Flg Off Murray Peden (*centre*) is flanked by Flt Sgt A. Stanley, wireless operator, and John Waters, bomb-aimer. (M. Peden)

down high-explosive and incendiaries around the Stirlings on their bomb-runs thousands of feet below.

Flg Off Murray Peden RCAF joined No 214 Squadron at Chedburgh as a pilot in September 1943 and completed a tour of operations on Stirlings and later Fortresses. He describes below how the fearsome imagery of night-bombing and the almost overpowering sense of self-preservation felt by the RAF's bomber crews while over the target are difficult to comprehend for those who were not there.

'One cannot write objectively and somehow contrive to gloss over the Stirling's poor altitude performance, although the rueful point may be made that our Stirling bomb-aimers were certainly poised much closer over their targets when they pressed the switch than our Lanc and Hally confrères above! What is also undeniable is that our curtailed altitude gave us a fearsomely close look at the targets Bomber Command was pounding so heavily. Nothing any film studio anywhere has done by way of attempting to portray the spectacle of our bombing attacks comes close to reproducing the stupefying reality. A modern filmmaker, transported back in time to the cockpit of a Stirling on a bombing run over a major German city, would, I am certain, be utterly dumbfounded. Scenes are engraved in my memory that a hundred years could not tone down, let alone erase, for in the late autumn of 1943 the bombing offensive against German industry – and our losses – had reached heroic proportions.

'We were quickly introduced to the defences of Hanover, Mannheim, Kassel and Bremen, and on the nights our crew rested, other Stirlings hit Frankfurt, Ludwigshafen and, on 22 November, their last German target, Berlin.

'Flying over such targets' defences at our puny altitude took us into a world that bordered on the surreal. High in the sky above us multitudes of chandelier flares descended in rows, swinging on their parachutes and lighting the embattled arena as though it had been placed under the canopy covering a 5-mile billiard table. Below us, great expanses of the city were wrapped in angry, darting flames, while others glittered with the sparkling incandescence that bespoke fresh tons of newly flaring incendiaries. Broadcast all through this maelstrom, flashes and smoky shockwaves afforded abundant though mute evidence of the torrent of bomb bursts striking home. At eye level and above, the sky all around us heaved, and occasionally bellowed, in the staccato accents of vicious flak bursts. In the fitful glare of this scene from Dante, one had a strong tendency to breathe only at lengthy intervals.

'With the clarity of opening a flagged photo album, I can still see and remember in the frightening glare 2 miles above burning Mannheim, the startling sight of a stricken Ju 88, passing us close by on a reciprocal course. In no more than two or three seconds, somehow protracted and magnified by the

A 4,000lb HC 'Cookie' explodes on Gelsenkirchen in the Ruhr Valley during a raid on 12/13 June 1944. The small white puffs are flak bursts. (via Chaz Bowyer)

strain and the devilish outlandishness of the smoke-clouded scene, that picture was burned into my mind ineradicably, complete with the detail of the German pilot's body-language as he bent forward to nurse his flaming, starkly black-crossed plane out of the centre of the flak-dotted arena, clutching for the sanctuary of blackness a few miles beyond the terrifying glare.

'It has become clear to me over the years that the intensity of the strain that prevailed on operations is a factor that non-participants simply cannot grasp. A writer once posed to another airman the question of what thoughts aircrews would have had, surveying that flame-drenched scene of hyperbolic violence sliding under their wings. I have forgotten what answer he made, but if the question had been put to me, I would have said, truthfully, that I can remember my thoughts and reactions basically focused at such moments on two, and only two, themes.

'The first was concentrating with fierce intensity on guiding the aircraft, at long last, over the shimmering green Target Indicators, and as precisely as possible – all the while poised to take some sort of makeshift evasive action if a fighter leapt at us – but listening intently while I was guided over this preview of Hell by the tense, ritualistic comments of the bomb-aimer, supine over his bombsight in the nose below me. All through the agonizingly slow bomb-run, defying survival logic and flying absolutely straight and level, we were terribly aware that we were, to the "enth" degree, sitting ducks, insanely vulnerable to flak and fighters, perched atop our ponderous cargo of high explosive, incendiaries and aviation gasoline, and lumbering forward with the imported unmanoeuvrability of open bomb doors. Firmly suppressed

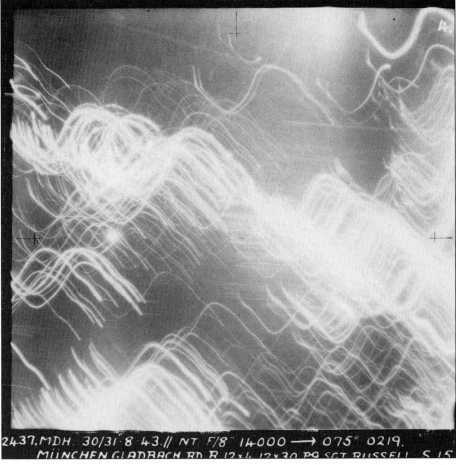

Fire tracks of burning incendiary bombs on the ground as seen from 14,000ft over München Gladbach by a Stirling of No 15 Squadron on 30/31 August 1943. (via Chaz Bowyer)

thoughts of half-a-dozen lethal hazards lurked near the surface throughout the glacially slow progress and nervous flagellation of the bomb-run.

'Our determination to carry out the run properly was not simply the result of our training; it was whetted by the fact that in most cases it had taken us two, three or four tense hours to reach this point, which made for a powerful desire to make this culminating effort worthy of its preceding ordeal, a desire rooted in the knowledge imparted to us at briefing of the target's continuing industrial contribution to the German war effort. Thus, pilot and bomb-aimer strained, upon entering the daunting "area of illumination", to select the best concentration of Target Indicators, and to team up to drop the bomb-load precisely at their centrepoint.

'Then there was the second force at work within us. Fighting for supremacy with that first, disciplined mental thrust, was a suppressed but powerful urge, which moved up to shove its predecessor abruptly aside the moment the bombs had gone and the camera had then methodically run off its five or six frames of target photos. These, too, forced suppression of nature, for they were of paramount importance, photographs synchronized to bracket the million-candlepower illumination of the photo flash that accompanied the bombs. This second force was the deeply ingrained instinct to survive, to escape this fiery scene of omnipresent destruction and strain toward the greater safety of the blackness that stood several tantalizing minutes away from the glittering Target Indicators below us. Keeping the survival instinct in second place in these surroundings took discipline, often all the discipline one could muster.

'As I recall, the writer earlier alluded to, commented in apparent surprise that the aircrew seemed to have given little thought to the fate of those below who might not be in safe shelters when the aerial deluge struck. My answer would have been to confirm his impression, and in crystal-clear terms. I certainly did not think about their fate; what I thought about exclusively the whole time the camera was rolling over was preserving one set of lives: my crew's and my own. I would have asked that writer to join us in the real world, or if he liked fantasy, to visualize himself being led towards the electric chair. Would his thoughts in the face of imminent death stray altruistically to the welfare of his executioners and their kin, or would they focus elsewhere with the concentration suggested long ago by Dr Johnson?

'I do not think anyone who knew me then would have classified me as insensitive – but I rise to remark, and with heavy emphasis, that when you are hanging in the air on a bomb-run, amid fighters, searchlights and heavy flak, unless you are earmarked for sainthood, you are likely to give serious thought to the precariousness of only one group of lives during that interval. Those were the lives held most precious, the ones that got priority. People who purport to have difficulty understanding that unknowingly reveal the highly pertinent fact that they have not personally had a good look at the "Old Man with the Scythe" in action, I mean really close up. In circles where two and two make four, however, I speak without fear of contradiction on his compelling persuasive powers.'

The incredible strain of operations was naturally best recognized by the aircrew themselves and on a raid they made allowances instinctively for the rare minor aberrations in patter and procedure to which even the best regulated nerves were prone in unexpected emergencies. Such lapses were almost always punished by humour in later re-enactments. Humour was the great and essential escape valve utilized by aircrew to bleed off some of the pressure. Peden continues:

'One of the best examples I ever heard of this humorous chastisement being administered was presented to me only a few weeks after I joined the squadron by a 214 pilot named Tommy Thompson. Tommy was an Australian and to appreciate his remarkable saga fully, you have to imagine hearing his pleasing twang superimposed on the laconic prose he was always careful to select in describing the event. Tommy's tale is worth repeating.

'He and his crew had been despatched in their Stirling to bomb a target well inside Germany – Hanover, if memory serves me correctly. Because of the routing, the trip was going to be longer than the usual five or five-and-a-half hours, but because the brass at Group HQ understandably wanted to hit the target as hard as possible, the bomb-load was not being correspondingly reduced. Complying with the austere laws of physics meant if the Stirlings were to get airborne at all, that something had to be shaved, and as Tommy languidly pointed out the chosen item was the fuel load. This was clearly pointed out to all pilots and flight engineers at the briefing, although without voting privileges, and not only pointed out but emphasized, with substantial redundant comment on the benefits to be derived from lean mixtures and all complementary features of prudent engine-handling.

'To complete the picture, a quick reference was made at briefing to the availability of Suffolk's emergency aerodrome, the inordinately spacious runways at Woodbridge, should prodigal throttle-wielders find it expedient to shorten the final leg for the purpose of ensuring a landing on concrete instead of in salt water or the bush.

'Tommy sketched in these salient features of the scenario, then added one or two final fillips. Fate had ordained that he would make this trip assisted, not by his own flight engineer, but with the services of the squadron's spare – nicknamed Bonzo – despite which "moniker" a gentleman highly qualified, experienced, and keen as mustard. On the down-side, Tommy had never

Flight engineer's petrol log for the Stirling I. The fuel states were recorded in the numbered columns (across the top) on an hourly basis and when a tank was turned on or off; a diagonal line was drawn when a tank was turned on or off while a reverse diagonal line was added to form a cross when the tank was empty, although in this example the engineer has written the word 'OFF' instead. The calculated tank contents were recorded after the diagonal line when each change was made.

flown with him before and thus the customary close rapport had to be built from scratch. Complicating matters even further was the fact that Tommy's regular aircraft was unavailable and he had been fobbed off with a hangar-queen spare bearing the reputation of being a greedy gas-guzzler.

'On the way to the target Tommy discovered that his "steed" had a further deficiency: it did not fly at all well. On this account he was forced to make an additional climb just as they approached the target, carrying him to the dizzy altitude of 12,000ft, the best they'd managed all night.

'The defences at Hanover lived up to their billing, Tommy pointed out: plentiful, active and frightening; thus he had his hands full until they dropped their bombs, obtained their photos, and clawed their way back beyond the baleful threat of the area of illumination. Up to this point he and Bonzo had little conversation beyond the occasional formal recital of a change of revs or boost, and a remark or two for the purpose of ensuring that such new settings were entered at the appropriate times in the engineer's log.

'Now that they were finally homeward-bound, Tommy was more than mildly interested in learning how they stood in terms of fuel still available, bearing in mind the extended route, the unplanned climb just before the target, and his aircraft's ugly reputation and performance. Affecting, for morale's sake, a nonchalance he was far from feeling, he addressed the spare engineer thus, in paternal Australian accents, "Bonzo, run over your log for a minute and see how we're doing for juice will you?"

'Bonzo snapped on the small red lamp above his knee and swished his pencil with professional speed up and down the columns of the log. In a matter of ten seconds he was reaching for the mike switch on the snout of his oxygen mask, and Tommy got the reassuring message, in strong positive tones: "We got bags of juice, Skipper."

'At this point in the story, Tommy looked at me with a gentle shrug and said apologetically, "I didn't want to hurt his feelings . . . tell him over the intercom he's chockfull of shit . . . but I knew we'd been using high revs and boost . . . and I'm remembering that extra climb – and climbs go at 500 gallons an hour. So I bite my tongue and decide I'll wait five minutes and ask him again."

'Five minutes later Tommy flipped on his own microphone and with a shade less casualness intoned, "Just check that juice again for us, will you, Bonzo?"

'On went the little red lamp again. This time it stayed on longer, considerably longer. In Tommy's judiciously exercised peripheral vision the pencil was seen to make, in Bonzo's flying fingers, repeated round trips over the log, and to retrace its steps at ever higher speed. Now Bonzo's hand flew to his mask and the mike switch clicked to presage a new message, this time resonant with frantic finality, "Skipper, we got fuck-all!"

'Tommy pursed his lips gently at this point in the story, turning face-on to me to reproduce his mild frown at these tidings, as he observed matter-of-factly, "Ya know, Murray. . . I was interested to know how we went from bags to fuck-all in five minutes."' It took me about the same length of time before I could stop laughing.

'The story went around the station with the speed of light and overnight became a standard gambit with all our flight engineers. The very next time I was flying with my own crew, and routinely asked mine to "check the juice", Bill Bailey came back in ringing tones, without so much as a glance at his log: "We got bags of juice, Skipper." After he got his laugh from the crew, and my jocularly profane rejoinder, he got down to business. The point was, Bonzo was a good engineer. But the best could make a slip under that grindingly heavy stress. I made my share.'

CHAPTER TEN

A Tale of Two Crews

On 7 November 1942 No 90 Squadron became the seventh RAF squadron to re-equip with the Stirling. The squadron reformed in Bomber Command's No 3 Group and flew Mk Is and IIIs from Ridgewell, Wratting Common (West Wickham) and Tuddenham. Nearly two years later, in June 1944, the last one was retired and the squadron finally converted to the Avro Lancaster.

The squadron made a significant contribution to the hard-fought Battle of the Ruhr, the devastating Hamburg fire raids and the famous attack on the Nazi rocket research establishment at Peenemünde, before it was finally withdrawn from front-line operations in November 1943, in common with all the other Stirling squadrons. No 90's war record with the Stirling stood at 111 bombing and resistance operations, and 100 minelaying ops, totalling 1,937 sorties for the loss of fifty-eight aircraft.

The following accounts from two of No 90 Squadron's aircrews represent a snapshot of Stirling operations over a period of one month in the spring of 1944. At this point in the war they were flying a mixture of Special Duties and short-range bombing operations from Tuddenham, a satellite of Mildenhall in Suffolk. Bill Burns was the flight engineer in Flt Sgt Jack Towers' crew, and WO Dennis Field was the captain of another. Of the two crew captains, only Field was to survive the war, four crewmembers (including Towers) were to die as the result of the operations described below, and one other was to receive the CGM for his bravery.

On the night of 25/26 March 1944 a mixed force of 192 aircraft, which included thirty-seven Stirlings, was despatched to bomb the railway target of Aulnoye in north-west France as part of the Allied plan to disrupt enemy communications in the build-up to D-Day. Flt Sgt Jack Towers and his men were one of those crews detailed to fly that night, as his flight engineer Bill Burns recalls:

'On 25 March we were returning from a bombing operation to Aulnoye when we found ourselves being "escorted" by a Ju 88 flying level with us on the starboard side. It made no attempt to attack us and at one point our mid-upper gunner fired a short burst in his direction, without result.

'Just after this the Ju 88 crossed beneath us and there was a loud bang. The kite gave a bit of a lurch and we thought we'd been hit. The skipper, Flt Sgt Jack Towers, sent me down the fuselage to assess any damage we may have suffered. Meanwhile, the Ju 88 flew away in a steep dive to starboard and disappeared into the night sky.

'About three-quarters of the way down the fuselage I almost put my foot through a large hole on the starboard side, close to the floor. We could only assume that it had been caused by the Ju 88 clipping us with his tail as he passed underneath and we considered ourselves very lucky.

'Two operations later on 11/12 April, we flew EH947:S on a Special Duties flight to drop supplies to the resistance in south-west France.'

In addition to the SOE sorties mounted by No 38 Group, No 90 Squadron had detailed twenty aircraft to join the force but two had been withdrawn prior to take-off and four were recalled.

'These trips were at low-level all the way, mainly in bright moonlight and usually of 8 hours or more. On the way home all sorts of small-arms fire would be encountered periodically. I suppose it was from bored gunners on sites off the beaten track who took a chance on loosing off at anything flying.

'We got a very near-miss under the starboard wing just approaching the French coast and within about 10 minutes the starboard-outer engine began to overheat. Soon after crossing the English coast the needle of the temperature gauge went off the clock and the plane began to rattle. The pilot attempted to feather the engine but to no avail and I went down the fuselage to manually operate the solenoid, but it made no difference.

'By this time we were over East Anglia and the engine was glowing when suddenly it burst into flame and the whole wing was soon alight. In what seemed like a split second the Stirling dived into the ground and by some miracle our skipper managed to flatten out as we hit the deck and the aircraft skidded across two fields, disintegrating as it went.

'The wings were torn off and the fuselage broken in two, the whole lot surrounded in flames. On impact I was standing at the engineer's position and was thrown against the bulkhead where it sheared off, breaking a couple of my ribs. In the midst of this inferno our wireless op bent over me and said in a quiet voice, "Come along, Bill, I think you'd better get out, the aircraft is on fire!".'

Flt Sgt Jack Towers crash-landed EH947 at Icklingham village but the rear gunner, Sgt Jimmy Powell, who had left his turret and was sitting in the fuselage

Jack Towers and his No 90 Squadron crew at Tuddenham in March 1944. *Left to right*: Bill Burns, flight engineer, Andy Milligan, mid-upper gunner, Jack Towers, pilot, Jimmy Powell, rear gunner, Eric Webster, bomb-aimer, and kneeling, Jack Stimson RAAF, navigator. (B. Burns)

Austin Motors-built Stirling III, BK784:O, of No 90 Squadron, is seen here at Tuddenham in the spring of 1944. It completed forty-three ops with the squadron before eventually crashing after take-off at Chippenham Lodge, Suffolk, on 23 May 1944. (D. Field)

awaiting landing, was killed. His turret was found intact some 100yd from the wreckage; Jimmy's lifeless body was found a further 100yd away. After three days in the Station Sick Quarters Bill Burns was given one week's leave, plus walking stick, and on the 21st he returned to operational duties.

The squadron lost another Stirling (EF182:M) from the same operation that morning in a crash-landing at Friston. Its pilot was WO Dennis Field. After being badly shot-up by light flak, Field's flight engineer, Charlie Waller, was seriously injured and died later in hospital. Dennis Field recalls the momentous events that culminated in their crash-landing at Friston, and the award of the CGM to his wireless operator, Flt Sgt Eddie Durrans:

'Our regular aircraft, LK516, was unavailable so I took EF182:M instead and on getting airborne discovered that she was starboard wing heavy. Having no internal aileron trim it meant rather annoyingly keeping a slight pressure on the wheel to keep her steady. Our chosen route was similar to that we had previously used to take us to the area, pinpointing on the loops of the Loire with careful avoidance of Bourges and other flak areas, and then setting course for the target. We found it without difficulty and dropped our containers on site, as always feeling great respect for the courageous individuals on the ground who refused to submit.

'The return flight progressed equally well as we passed our crossing point on the Loire near to Blois, and I prepared to climb towards the coast. I avoided a train further up the line east of Le Mans as I built up speed and

Field's original crew when they joined No 90 Squadron: *front row, left to right*: Tony Faulconbridge, mid-upper gunner, Charlie Waller, flight engineer, Alan Turner, navigator, G. Royston, rear gunner, Arthur Borthwick, bomb-aimer, and Eddie Durrans, wireless operator. Of these six men, Royston went missing on 29 January 1944 flying as a spare gunner with another crew; Waller was killed on 11 April and Durrans was badly wounded on the same op, but received the CGM for his bravery. (D. Field)

then, past the railway, pulled up the nose. Suddenly, at about 2,000ft both Jim Blackwell and Tony Faulconbridge (rear and mid-upper gunners) shouted "Weave!", but as I rammed the aircraft down into a steep diving turn to port there were loud explosions just behind the cockpit as we were raked down the starboard side by light flak which poured at us from the train.

'I kept her down and after a short time the firing, and our gunners' reply, ceased. Alan Turner, the navigator, immediately came on the intercom saying that Charlie Waller, the flight engineer, had been badly hit in the head, trunk and legs, collapsing over his table, and that the wireless op Eddie Durrans, standing in the astrodome, had been severely wounded in the leg and also in the back. Jim and Tony reported safe and damage-free and I sent Arthur Borthwick, the bomb-aimer, back to help and report. I asked Alan for a course but his charts were covered and ruined by Charlie's blood and other body fluids. A chunk of shrapnel missed Alan's head by inches and shattered his Gee set – had he been any taller he would have been decapitated.

'I made a quick estimate and steered due north and then opened to full power, built up speed and started the climb to height. A superficial check indicated that structurally the engineer's station as well as Alan's was a shambles, the shrapnel having entered through his side panel, destroying many of the instruments, and a piece of metal had also damaged the pitch lever bracket. Charlie was in a bad way and had been laid on the floor and given a morphine injection, but he still tried to help with advice on controls even though he must have known his chances were slim. Emergency dressings were applied to Eddie's wounds and he then returned to his station refusing morphia in case it affected his reflexes, and continued to transmit and operate his set throughout the rest of the flight in spite of severe pain and shock. Whatever else had occurred I did not know, but the engines seemed OK although the fuel position and operation were uncertain because of the damage.

'After gaining height I maintained my assumed course with Arthur and Alan tending and doing their best for the wounded, and Tony and Jim on very full alert. After about half an hour we weaved over the coast and headed across the Channel. Eddie sent out distress calls which he cancelled as we neared the English coast. Expecting to see land at the earliest opportunity I was relieved to see a flarepath ahead on the edge of the cliffs. I called up on "Darky" but the R/T had been damaged and reception was poor. After several attempts I just distinguished permission to land and commenced the circuit after requesting urgent assistance on landing for the wounded.

'The flarepath leading out to sea looked rather short but I thought I could make it and was very anxious to get aid for Charlie and Eddie, apart from doubts about fuel. On the downwind leg I selected undercarriage down but nothing happened and Alan had to wind it laboriously down by hand, several hundred turns of a heavy lever, while I circled. The green lights finally came on and I came in low touching down on the edge of the grass runway. About halfway along the stretch of lights ahead was rapidly running out and I rammed open the throttles.

'We were barely above stalling speed as we shot over the cliff edge and I was fortunate to sink 50 to 100ft and gain flying speed for a climb but unable to retract the undercarriage. The second approach and touchdown at minimum speed were similar but this time there was insufficient room for overshoot. I made the instant decision to swing off, applying full rudder and brake. The port wing dipped as the aircraft careered round and then the undercarriage collapsed. We thumped to a jarring halt, clods of torn earth flying up from the distorted propellers until I switched them off. No one had been further harmed by the violent arrival and there was no fire.

Tony Faulconbridge, mid-upper gunner in Dennis Field's crew, pictured in the turret of a Stirling at Tuddenham in 1944. (D. Field)

'Charlie lay still on the floor and was too injured to be moved inexpertly except *in extremis*. When I clambered out of the overhead hatch the others were helping Eddie from the rear escape. We carefully lifted him down, moved him clear and made him as comfortable as possible. By now the ambulance and fire engine should have arrived but there was no sign of them so I clambered back over the wing into the cockpit and called up on the R/T requesting in unequivocal terms that they pull their fingers out. They arrived shortly afterwards. The MO gave further first aid and the stretcher party carefully carried Charlie out through the rear door and into the ambulance to join Eddie for the short ride to Eastbourne hospital. The rest of us went to sick quarters for a quick check-up.

'We learned that we had landed at a fighter station called Friston adjacent to Beachy Head, which explained the short grass runway, a point which they had unsuccessfully tried to warn me about but which I had not heard because of the poor reception. Thence to a hastily alerted intelligence officer for debriefing. He was a little frustrated when I told him I could say nothing except to inform Tuddenham of our situation and that the operation had been successful. Then we departed for a meal and fitful sleep.

'Next morning we paid a visit to the hospital. Our fears were realized when we were told that Charlie had died during the night despite massive

Field (*fourth from left, standing*) pictured with his crew, groundcrew and their regular Stirling III, LK516:J, at Tuddenham in May 1944. (D. Field)

blood transfusions. Eddie had successfully undergone surgery to remove shrapnel from his leg and back.

'We left the hospital in sombre mood and went back to Friston to recover any important movables left in the aircraft. The collapsed kite was near to a similarly foundered Halifax which I thought for a moment I had written off as well. But we had fortuitously just missed it. Burdened with maps, charts, and paraphernalia and having obtained rail passes, the five of us were given a lift to Eastbourne station to catch the train to London. It was too late for connections back to base and so, pooling the small amounts of money we had, we were able to pay for a meal and night's accommodation at a forces hostel. Next day we travelled across London on tube and bus and returned via Cambridge to Tuddenham. It was perhaps typical of the times that five scruffy, hatless unshaven aircrew carrying parachutes and helmets and one with his tunic covered in dried blood excited not a glance or comment, let alone an offer of a lift or assistance.'

Eddie Durrans was transferred to the RAF Hospital at Halton in Buckinghamshire where, on 25 April, he was informed of his award of the

CGM. But not until the following year had he sufficiently recovered from his injuries to attend an investiture at Buckingham Palace on 5 March 1945.

Little more than a fortnight later Bill Burns and his crew were involved in yet another crashlanding, once again with fatal results for one of the crew:

'The very next operation was another SOE job on 28/29 April in EE974:W and almost exactly the same thing happened again – same engine, too – but this time we were at 3–4,000ft on the approach to land when the engine caught fire. All efforts with fire extinguishers and feathering procedures were in vain and the skipper gave the order to bale out while he did his best to hold up the dead wing.

'I was the last out at about 1,000ft and hit the deck immediately the parachute opened. I gathered up my parachute, draped it around my shoulders and set off in the early dawn (05.30hrs) across a couple of fields towards a distant farmhouse. My access to the farmyard was barred by two barking dogs, chained one each side of the gate cutting me off. The barking roused the farmer who opened an upstairs window and called down, "What d'yew warnt?" in a rich Suffolk brogue.

"Can you tell me how to get to the nearest RAF station?" I enquired.

"D'yew warnt a cuppa tay?"

"No thank you," I replied. "Just tell me where the RAF is."

"I'm now gonna mek one," he insisted. Again I politely refused.

"Yew jis stay thar," he called. In a moment or two he had come down.

"Where d'yew cum from then?" he enquired.

"I just came out of that aeroplane you can see burning across there." I gestured over the fields to where my aircraft had crashed about ½ mile ahead.

"Oh," he said, followed by a long pause. Then, pushing my chute to one side, he gave me a scornful look and said disappointedly, "RAF! I thort ee were a Jerry!" Then he leaned on the gate and said into mid-air, as if I wasn't there, "Not a bad mornin' is it?".

'I suppose anyone could give an RAF bloke a cup of tea as his farm was near Stradishall, but to give a Jerry one – well, that would have been a tale worth telling.'

EE974 crashed 2 miles north-east of Stradishall. Its pilot, Flt Sgt Jack Towers, never got out of the burning aircraft and was killed – ironically, on his 29th birthday. His rear gunner, Sgt Andy Milligan, almost made it. Very much alive, he baled out but nearing the ground his parachute 'hung up' in a large tree. He released his harness but broke his neck on hitting the ground and died instantly.

Following these two consecutive accidents Bill Burns and his bomb-aimer, Eric Webster, were declared 'Temporarily Unfit For Flying Duties' for some time afterwards. Plt Off Entwhistle, the crew's Kiwi wireless operator, transferred to No 75 (NZ) Squadron shortly afterwards and perished on the first op with his new crew.

As for No 90 Squadron itself, it became the 43rd squadron to convert to the Avro Lancaster, receiving its first aircraft in May 1944. But its final operation with the Stirling was not flown until 17 June when the squadron attacked a rail cutting south of Montdidier with a mixed force of Stirlings and Lancasters, thereby concluding eighteen months of service with the type.

CHAPTER ELEVEN

Most Secret

The opening months of 1944 saw four of Bomber Command's Stirling squadrons temporarily supporting the efforts of Tempsford's two hard-pressed Special Duties squadrons by dropping personnel, weapons and equipment to Resistance groups in occupied Europe and Scandinavia. They were soon joined in this endeavour by No 38 Group's squadrons, operating the Stirling IV, who continued to fly in this role until the end of the war in addition to their primary tasking as paratroop and glider transports.

Flt Lt Bob Chappell and his No 149 Squadron crew were typical of Bomber Command's contribution to Special Duties operations in 1944, which they flew

Plt Off David Mitchell, navigator with Flt Lt Bob Chappell's No 149 Squadron crew, pictured at Methwold in June 1944. (D. Mitchell)

Tempsford was the wartime home of the RAF's two dedicated Special Duties squadrons: No 138 flew Stirling Mk IVs from June 1944 and No 161 flew Mk IVs from September 1944. On the blackboard in flying control a WAAF chalks up the station's latest arrival home following the night's Special Duties operations on 11 April 1944. Although the station was not operating Stirlings at this time, the rarity of photographs depicting operations at Tempsford has warranted its inclusion here. (C. Annis via K. Merrick)

in addition to other sorties such as minelaying and precision bombing of strategic targets in France. Between September 1943 and September 1944 they completed forty sorties of which fifteen were Special Operations Executive (SOE) 'parachutages' trips to the French Resistance.

Plt Off David Mitchell was the navigator in Bob Chappell's crew and he relates below something of their experiences on Special Operations:

'Our whole tour was fairly uneventful although we lost a lot of aircraft and friends on the squadron. So we flew with a constant feeling of wondering when our turn would come, a strain which was often diluted by plenty of beer and mad sing-song parties in the mess during stand-downs. But for the most part we had no trouble, apart from icing and some hair-raising moments flying round mountain peaks in the Alps looking for our dropping sites.

'Supplying the Resistance was all low-level work, flying at 500ft across France, identifying our dropping site by a signal from a solitary figure in some remote field or plateau using a lamp or torch. Once codes were exchanged successfully three more lights would spring up in a line which identified the wind direction. We would make the drop from 150–200ft flying into the wind.

'We were fortunate enough to find all our reception parties and, having exchanged correct signals to drop our canisters, and make our way safely home. Except once. The fact that we did not have our own pilot that night has got nothing to do with the events that followed.

'It was the night of 10 April 1944. We had to fly with a substitute pilot, Flg Off Alan Bettles, and it had taken us a good four hours at 500ft to reach the dropping zone most of the way across France. Arriving bang on ETA, sure enough as always the reception was there. The three lights were already on, indicating the wind direction, but on this occasion they were signalling the wrong letter(s). There was mild panic while I hurriedly checked my charts, the chosen pin-point and DR run. Sure enough, I confirmed to the pilot that this was for certain our dropping site. We made several circuits at a cautious 300ft but we still did not get the pre-arranged signal code which would have allowed us to make the drop.

'The frustration was heightened when once again I became the focus of attention. Pilot to navigator: "Are you really certain this is our spot?"

At this point I was beginning to doubt myself, especially since we noticed several dropping zones on the way. Many would give you a preliminary flash, hopefully. So I hurriedly rechecked everything once again. Navigator to pilot: "This is most definitely our dropping point. Time is getting on. If they don't give us the correct letters we're getting out of here!" This was a terribly frustrating decision to take and so reluctantly we started to climb and make our way home.

'"They're doing it right now!" It was our rear gunner "China" Town shouting into his mike – "They're doing it right now, they're giving us the correct letters!" And he was right! So round we circled to make another approach. Bomb-doors open. Wheels partly down and flaps partly down to reduce speed. We made our dropping approach. 200ft . . . 150ft . . . Suddenly all hell let loose. Two searchlights opened up straight on to us. We seemed to be under a lot of fire from at least three guns on the ground, at point-blank range.

'The bomb-aimer, George Mackie, was in his niche beside the bomb-sight. The flight engineer, a young, canny, over-conscientious Scot named Ian Harvie, was a little concerned about the amount of time we had spent in the target area, and was making a quick check of the fuel gauges. So, there was no one in the second pilot's seat and the pilot was screaming his head off to this

Bob Chappell's crew. *Back row, left to right*: Ian Harvie, flight engineer, Bob Chappell, pilot, David 'Mitch' Mitchell, navigator, George 'China' Town, rear gunner. *Front row*: Gordon 'Taffy' Thomas, wireless operator, Harry Foxton, mid-upper gunner, George Mackie, bomb-aimer. (D. Mitchell)

no one – "For Christ's sake give me more boost, more revs!". He was weaving as much as he could with both hands on the stick.

'It was Taffy Thomas, the WOp, who obliged. He rushed forward from his table knocking everything flying and bruising himself black and blue on the way and pushed the throttle levers forward – as he described it, "pushing everything through the bloody gate". The good old Hercules engines responded with a roar. For a moment we thought they had jumped the aircraft, like horses at the starting gate.

'During all this time "China", the rear gunner, was taking care of the searchlights with a few accurate bursts from his guns. He said later he was reluctant to fire on the scurrying figures below who normally he knew to be our "friendly" resistants.

'I'll say this for our pilot Bettles, he wasted no time. We were out of that danger zone like a flash. And so we made our way home, the first time we had not delivered the goods. To relieve our disappointment and frustrations

No 149 Squadron's R-Robert, skippered by Flt Lt Bob Chappell, formates at 200ft over the Suffolk countryside on 14 July 1944 with two other squadron Stirlings, practising for an SOE operation. (via D. Mitchell)

we shot-up two trains and some road transport on the way home, much to the flight engineer's horror. He was still checking those fuel gauges. The total flight took 8hrs 40mins, so we didn't have a lot of juice left. When we landed none of us was the least bit tired.'

Plt Off Den Hardwick and his crew joined No 299 Squadron in No 38 Group on 3 March 1944, following a tour of bomber operations with No 149 Squadron at Lakenheath. They were stationed at Keevil in Wiltshire and flew a number of SOE sorties deep into occupied France in the months following D-Day. Den recalls in particular the night of 15 September when they took a team of 14 SAS paratroops to a drop zone close to Strasbourg on the River Rhine:

'The weather was good until we came to the Rhine valley where visibility on

Fresh out of HCU, Den Hardwick and his crew pose for the station photographer at Lakenheath on joining No 149 Squadron in October 1943. *Back row, left to right*: Sgt Tom White, flight engineer, Flt Sgt Gordon McCleod, air gunner, Flt Sgt Ted Webb, navigator. *Front row*, Flt Sgt Les Fahy, wireless operator, Flt Sgt Den Hardwick, pilot, and Flt Sgt 'Ketchy' Ketcheson RCAF, bomb-aimer. (D. Hardwick)

the ground was nil due to dense fog. I called the stick leader, an SAS captain, up to the front and showed him the view from the cockpit. His only question was, "How close to the DZ can you put us down?". After discussion with Ted Webb, my navigator, we calculated that with a direct run from where we were, based on our last positive fix, the worst error would be 20 miles, to which came the reply, "We can walk that tonight and they won't see us in the fog." We made our run and away went the lads.

'Some few weeks later after 299 had moved to Wethersfield, the crew went for a night out in Chelmsford. We walked into a pub and there, believe it or not, having a beer were the lads we had last seen jumping into the fog near Strasbourg. What a night that turned out to be!

'On the night of 31 August we set off on an SOE trip to southern France, somewhere north of the Pyrenees. We were flying in a virtually brand new Stirling IV, LJ971, and this was our third trip in her. Around the DZ we lost one engine; halfway across France on the return trip another engine started to overheat. By the time we reached northern France we were making very slow progress and gradually losing height, and it seemed there was no way we would make the Channel crossing. After D-Day the navigators used to draw a red line on the charts showing the latest information as to where the front-line was. It was a bright moonlit night and we spotted an airstrip which, according to the chart, was on our side of the front-line. We managed a two-and-a-half engine landing on a runway with no lights which was quite hair-raising.'

Tom White was the flight engineer in Den Hardwick's crew and he remembers the incident vividly:

'The sound of gunfire several miles away and the sight of all the mines that the Germans had left lying around, convinced us that this was no place for us to be. The Army took us to a tent and fed us on steak and boiled potatoes. We suspected there must have been a three-legged bullock in the vicinity.

'In the morning we found that another Stirling – from No 196 Squadron – had also landed at B17 [Carpiquet airfield, west of Caen], and its skipper was Henry "Chuck" Hoysted. His flight engineer and I decided that the engines of Hoysted's aircraft, ZO-D, were serviceable so they gave us a lift home to Keevil.'

Above and overleaf: Shot up and badly damaged by a Junkers Ju 88 nightfighter following an SOE drop over Denmark on 28 September 1944, No 138 Squadron's Stirling IV, LJ932:N, limped home across the North Sea on one engine to make a miraculous crashlanding at Ludford Magna near Louth, Lincolnshire, at 02.35hrs. The damage inflicted during the attack on the Stirling's port mainplane, elevator and port inner engine can be clearly seen in these photographs.

Despite sustaining serious injuries to his arms, legs and buttocks as a result of the fighter attack, Plt Off Sam Curtis, the flight engineer, bravely assisted his pilot, Flt Lt Read, and the wireless operator to coax the badly damaged Stirling homewards. For their bravery each was awarded an immediate DFC. (Mrs N. Curtis)

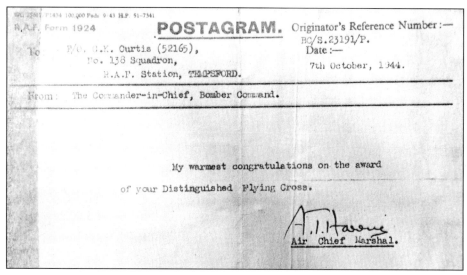

The personal telegram sent to Sam Curtis from 'Butch' Harris, C-in-C RAF Bomber Command, congratulating him on his award of the DFC. (Mrs N. Curtis)

Another Keevil-based crew to fly SOE supply drops during the summer of 1944 was that of Flg Off Gib Goucher RCAF, also of No 299 Squadron. One particular sortie caused them a few problems as Goucher's flight engineer, WO Leonard Brock, recalls:

'During August of 1944 we carried out many night sorties over France but one was quite significant. On 2 August we took off in Stirling LJ919 on a night drop over France called "Horace 7" with twenty-four containers and one pannier. We met heavy flak over the DZ and returned to the UK, but we were diverted to Weston Zoyland in Somerset because our starboard elevator had been damaged by the flak. We made a successful landing and left the aircraft behind to be repaired by groundcrew, and flew back to Keevil.

'We returned to Weston Zoyland on 4 August, by which time the aircraft had been repaired, and we found that all fuel tanks had been filled to capacity. We could take off all right but we could not land as we were above the all-up weight for landing. We asked permission to drop the containers but this was refused and we were told to fly over Salisbury Plain and jettison fuel from the main tanks. We dropped 1,170 gallons of 100 octane by opening valves inside the aircraft and the petrol went out in a thick black swirl. We managed to land back at Keevil and took the load back to France the following night and this time dropped it in the right position without opposition.'

CHAPTER TWELVE

The Dawn of Liberation

Stirlings of No 38 Group were in the vanguard of Operation 'Tonga' on 5 June 1944, the eve of D-Day, to drop paratroops of the British 6th Airborne Division in Normandy. Their task was to secure the left flank of the Allied landings due to take place on the morning of 6 June. They followed up this success in the early evening of the 6th by towing Horsa gliders across the Channel to Normandy carrying troop reinforcements to consolidate the bridgehead.

Activity both in the air and on the ground was intense, as witnessed by the air- and groundcrews of Nos 196 and 299 Squadrons stationed at Keevil in Wiltshire. Fred Baker, an engine fitter with 'A' Flight, No 299 Squadron, recalls the climax of the busy weeks leading to D-Day:

Stirling IVs of No 196 Squadron line the runway at Keevil prior to Operation 'Tonga', 5 June 1944. (IWM CH21187)

Den Hardwick's No 299 Squadron Stirling IV, EF267, 'The Saint', pictured on its dispersal at Keevil in June 1944. (D. Hardwick)

'At last the big day came. On D-Day minus one, 5 June, we checked our kites and ground-tested them ready for the aircrews to marshal the aircraft on to the runway. As the day went on, in came the Paras who eventually boarded the aircraft after writing all sorts of slogans on the fuselages. It was at 23.00hrs that they began to take off. The doors of the aircraft were still open and you could hear the lads singing *Shoo Shoo Baby*, a popular song at the time.'

Cpl Jack Parker, a wireless mechanic with No 196 Squadron, remembers: 'On the evening of 5 June it was strange to see around the airfield perimeter many of those trades whose duties usually kept them on the main domestic bases – clerks, cooks, stores.'

Keevil's two squadrons were part of a larger airlift tasked with delivering paratroops of the 6th Airborne Division's 5th Parachute Brigade to the eastern banks of the River Orne in Phase 1 of Operation 'Tonga' on the eve of D-Day, 5/6 June. The airborne troops were to secure the Allied left flank in the opening of the invasion of France. Plt Off Den Hardwick and his No 299 Squadron crew (see Chapter 11) flew to France in Stirling IV EF267, 'The Saint', carrying twenty troops from 225 Parachute Field Ambulance, Royal Army Medical Corps, and nine equipment containers, under the command of Captain George Holland. They were to take off from Keevil at 23.55hrs and Hardwick was to drop the troops at 00.50hrs on 6 June at DZ 'N', ¾ mile east of the River Orne:

'On the night of 5/6 June we took off for France taking Captain George Holland and a medical team. Shortly after take-off the Gee equipment went on the blink and as we approached the Channel it packed up altogether. The weather wasn't all it might have been with quite a lot of low cloud. As we came up to the French coast the most positive landmark we could see was Le Havre. We decided to make this point our landfall and make a direct run from there to the DZ where we got rather a warm reception from the defences but got away with it. Three or four weeks after the drop I received a letter from George Holland thanking us for dropping them on the right spot. All had landed safely and we were all very touched that he had found the time to drop us a line in what must have been hectic days just after D-Day.'

Flg Off Gib Goucher's No 299 Squadron crew took off from Keevil at 23.35hrs on 5 June in Stirling EF243:S. Goucher's flight engineer, WO Leonard Brock recalls:

'We met heavy flak over the DZ and I was wounded in the left leg but we did drop our paratroopers in the right place as I was to learn later. We had both tyres punctured by flak and cannon fire and had some 150 holes in the aircraft and a hole in the port tailplane, but we landed safely on the rims of the wheels and pulled the undercarriage up on touching down. This shows what punishment a Stirling could take. I was admitted to hospital at Melksham for treatment to my injured leg. I did not return to operational duties until 23 July.'

Flt Lt George Copeman, a wireless operator with Sqn Ldr D.W. Triptree's No 299 Squadron crew, remembers the close teamwork between air- and groundcrews that made Keevil's contribution to D-Day possible: 'We dropped the paratroops in Normandy at about 02.00hrs and returned and tried to sleep, while the groundcrews worked hard to refuel all the aircraft and line them up, closely packed, at the eastern end of the main runway, with tow ropes attached to the Horsa gliders.'

Fred Baker and the rest of the squadron groundcrews were kept very busy preparing the aircraft for the drop later that evening of troops from the 6th Air Landing Brigade:

'After the aircraft were clear of the drome we went back to our flight dispersal to fill empty petrol cans with engine oil ready for the return of the aircraft for

225 PARACHUTE FIELD AMBULANCE, R.A.M.C.

B.W.E.F.
6 July 44.

Dear Hardwick,

Just a few lines to let you know since I last saw you that all is well. Thanks for your despatch, we are alive & kicking, and all most grateful to you & your crew for putting us down in good shape safely & putting us down in midnight M.S.T. The whole show was excellent, and I must confess I am L- flying to get out of this kit. Seeing most flak coming up when too lots's were open into an awe-inspiring, not to say, sobering sight. I expect it was a long way away but it looked too damn near to me. I hope you had a good trip back.

P.S. I suppose you have been over here doing the supply drops. Feel like climbing up in there & giving "Timm" a lift back. When I see you will go over & give my regards to Steve & Harry, & no doubt by the story we shall have a party when we come back in due time. So you had plenty of the Trimmin'. That the J & the things S Herries round here — Bees. to stand drink it will all when we got who. well worth it all when we got who. Well many thanks to you & your crew for the safe & comfortable transportation of us, a very fine drop. Hope to see you soon.

Yours,
George Holland
Capt., RAMC

A copy of the letter received by Den Hardwick from Captain George Holland RAMC, written one month after he and his field medical unit were successfully dropped from Hardwick's Stirling over Normandy in the early hours of 6 June. (D. Hardwick)

Aussie crews with No 196 Squadron pose for the camera at Keevil in June 1944. *Back row, left to right*: Ron de Minchin, his bomb-aimer, Smith, Chuck Hoysted, McClaren, McCarthy, Mann, Tickner. *Front row*: Caldwell, Light, Marshall, two WOp/AGs, Keith Prowd. (via K. Prowd)

refuelling. Having only one 450-gallon oil bowser it would have been a big job topping up the oil tanks on the aircraft, so we formed a chain up through the fuselage, through the hatch and on to the mainplane. In the meantime the petrol bowsers were going from plane to plane topping up the fuel tanks while other lads were clearing out the sick bags. It was a very quick turn-around and we got the aircraft marshalled back on the runway ready to take off again with the Horsa gliders that evening.'

The activity on 5/6 June was equally intense at the other Stirling airfields scattered across southern England. Sqn Ldr Ray Glass, 'D' Flight Commander on No 1665 HCU, recalls: 'On D-Day, 1665 HCU had sixty trained reserve crews on stand-by. We were invited to watch one of the front-line squadrons taking off with gliders and troops. One of their Stirling aircraft went U/S and the troops in the glider were furious and clamoured to join their colleagues, but to no avail. The remaining Horsas were all fully loaded and were towed off without them.'

Built almost entirely from wood, the Airspeed Horsa glider was capable of carrying twenty-five to twenty-eight fully equipped troops and two glider pilots. Horsas were towed by Stirlings to Normandy and Arnhem during 1944, and on the Rhine crossing in 1945. (MAP)

One groundcrewman remembers the eve of D-Day '. . . as one of great activity with the preparation of aircraft and gliders, ground-testing and loading, followed by the arrival of the paratroops. The day itself was just hell with the departure of fully laden aircraft and gliders. On their return, some damaged, we had to repair and service the aircraft and re-load them with bales of tyres and wicker hampers filled with rolls of barbed wire. These items were just dumped out of the planes over France without chutes. The whole episode was one never to be forgotten.'

Shortly after 19.00hrs on the evening of D-Day, thirty-three Stirling crews from Keevil flew back to France for their second trip across the Channel in less than 24 hours, towing Horsas containing Divisional Troops of the 2nd Ox & Bucks Light Infantry. The first of the Horsas began to land in France on LZ 'W', west of the Caen Canal, at 21.30hrs. Flying in Stirling 'G' skippered by Sqn Ldr D.W. Triptree, George Copeman recalls:

'We took off in the early evening of the 6th and formed up for the flight over the Channel, flying at a steady 1,000ft. After casting off our glider, those who could watched intently as the gliders dived steeply down and smashed their way through the poles on the landing zone. There was more heavy machine-gun fire than during the previous night when the tracer looked pretty, but it was very effective.

'It was a magnificent sight with ships everywhere, landing craft, aircraft, and HMS *Warspite* firing its 16-inch guns over the top of our troops, big black belches of smoke with each salvo. When we turned and went out to sea we passed near the *Warspite*. Shortly afterwards Tosh Truelove, our tail gunner, happened to mention one of our Stirlings hitting the water near the *Warspite* and I knew instantly that it was Bob Clark's aircraft. As I looked back it was already half submerged and burning furiously. Nothing could be identified.

'When we got back, Clark's aircraft was certainly missing but how could I have known it was his aircraft near the *Warspite*? Some weeks later when the 299 Squadron adjutant, Roy Fischel, was on leave, I was sitting at his desk trying to keep his job under control, when in came a letter from the Admiralty. It said that on D-Day a Stirling had hit the water near the *Warspite*, which had put out a boat and picked up the body of a Flg Off Boyce. He was Clark's navigator.'

Flg Off J.H. Clark and his five-man crew in Stirling 'K' had successfully cast off their Horsa glider, piloted by Sgt Richardson of the Glider Pilot Regiment and carrying nine troops, but as Boyce gave Clark a course for home their Stirling IV was hit by flak and within minutes it had crashed into the sea in flames. No one survived the crash and only two bodies were recovered from the sea by the air-sea rescue service. They were later identified as Flg Off F.A. Boyce and WO E.H. Shrump.

However, unbeknown to George Copeman and his crew, this was part of a double tragedy that had unfolded around them that evening. The glider they had cast off over the LZ, piloted by Lt Martin and carrying eight fully equipped troops, had crashed on landing, killing the pilot.

CHAPTER THIRTEEN

Airbridge to Arnhem

Arnhem was the last major defeat in battle suffered by the British Army. Almost 12,000 men of the 1st British Airborne Division, the Polish Independent Brigade Group and various attached units took part in the operation which began on 17 September 1944; within nine days nearly 1,500 of their number had been killed and 6,500 taken prisoner. The rest were evacuated to safety.

Six Stirling squadrons of the RAF's No 38 (Airborne) Group and hundreds more Allied transport aircraft were involved in this daring, but ill-fated, operation to capture the bridge over the Rhine at Arnhem.

The view from the ground at one of the RAF stations involved in the maximum effort for Arnhem was one never to be forgotten. John Roberts, a Fleet Air Arm fitter seconded from RNAS Lee-on-Solent to RAF Keevil, remembers:

'Keevil airfield was closed and all 50 Stirlings and 150 Horsa gliders were painted with encircling black and white stripes, an operation that was completed in a single day! Awaiting favourable weather meant that the ten men on guard duty, including me, were on continuous duty for four days and five nights without sleep. Not surprisingly, we slept for a full 48 hours after the operation took off.

'But you should have seen the take-off: St Mary's Church in the village stood like Eros in the middle of Piccadilly Circus as the Stirlings and gliders set off on the historic flight.'

For the men of the Stirling squadrons who were to fly the operations over the coming days, many were to run the gauntlet of enemy flak and fighters over Holland in broad daylight and at low-level. Sgt Mike 'Taff' Stimson was wireless operator with Australian Chuck Hoysted's No 196 Squadron crew. His memories of their involvement in a re-supply flight on Tuesday 19 September, and of squadron politics, remain vivid to this day:

'When we were all lined up on the peri track at Keevil with our engines running, my skipper Henry "Chuck" Hoysted always liked me to sit down by the

'Chuck' Hoysted and crew, No 196 Squadron, 1944. *Left to right*: Flt Sgt Ray Owen, bomb-aimer, Flt Sgt Mike 'Taffy' Stimson, wireless operator, Flg Off 'Chuck' Hoysted DFC, RAAF, pilot, Flt Sgt Jack Hooker, air gunner, Flt Sgt John Barker, navigator. Kneeling in front is Flt Sgt Bill Garretts, the flight engineer. (M. Stimson)

The damage meted out by German ground defences at Arnhem to the RAF's Stirlings on re-supply drops kept the maintenance staff busy at home. Here a Stirling IV of No 196 Squadron receives attention from engineering staff in a hangar on the technical site at Keevil. (via A. Thomsett)

Paratroops of the 21st Independent Parachute Company prepare to embark in Stirlings of No 620 Squadron at Fairford, Gloucestershire, on the morning of 17 September 1944 – destination Arnhem. (IWM CL1154)

A Stirling tows a Horsa glider over the Dutch coast heading for Arnhem. (IWM CL1147)

back entrance door while we were taxiing. This was so I could lean out and look underneath the wings to tell him if there were any obstructions beneath that he couldn't see from the cockpit because of the nose being so high. I had to make sure that the main wheels stopped on the perimeter track. Right from our very first op dropping supplies to the Resistance, I found that my intercom lead wouldn't reach the plug in the aircraft, so I had it lengthened after the first operation by adding another intercom lead to make it some 6–8ft in length.

'So, I was sitting there on the taxi track when Jim Metcalfe came running up to us saying his pilot had set an engine on fire and could he come with us. I hurriedly told Chuck about it. "Get him in," he said, "I can always do with another pair of eyes up front. Tell him to come up front and sit in the second dicky's seat."

Glider pilot's eye view. This photograph was taken from the cockpit window of a Horsa glider as its Stirling tug (visible at the top of the picture) tows it over the Dutch coast *en route* to Arnhem on 17 September. (IWM BU1160)

'Jim came to Arnhem with us and returned safely. His navigator who went with WO Keith Prowd's crew got killed when they all had to bale out with two engines on fire, the third overheating and about to catch fire. They baled out so high that they were shot to pieces in their harnesses as they floated down. Keith saw four of his crew hanging by their harnesses in the trees all shot up. Lofty Matthews, flight engineer, was seen to come out on a pannier parachute, a smaller 'chute intended for containers. He either fell off or was shot off and killed instantly. (His wife was a Canadian nurse who'd travelled to the UK to be with him. She arrived at Steeple Ashton the day he was killed.)

'Our skipper, "Chuck", had been incensed by rumours passed around the station by the "Poms" about the Aussies hanging back when it came to facing the flak. Chuck was determined that they wouldn't say it about this Aussie. The result was that when we got to Arnhem there were only two people in front of us – the wing commander and the squadron leader. Earlier that day the wing commander had insulted us, calling us pigs, and had told us that we would fly at 500ft, no higher, no lower. Any crew that flew above or below this height, and his number was taken, would be out of aircrew.

'We had also been told to formate in threes as much as we could to increase our firepower. One Stirling was in front of us about 400yd away, if that. One minute there were three of us, the next minute the middle one had just gone "whoomph" into thousands of bits. We ran the risk of flying through the debris. Luckily the explosion had blown everything out and it was just like flying through a black hole and we emerged the other side virtually untouched, but the fellows on the other side got damaged by bits of the aircraft.

'Just as two more aircraft further in front of us began to drop their supplies and we were about to drop ours, some aircraft from another station that had been misbriefed came in from our port quarter at right angles, 1,000ft above us. Wicker baskets containing five cans of petrol each were dropped without parachutes from these aircraft, narrowly missing us. We dropped our supplies and Chuck put the Stirling into a sideslip, nearly taking the top off a church steeple. Above the noise of the engines, and although wearing a headset, I could still hear outside our aircraft the sound of "whoomp, whoomp, whoomp, whoomp". That was the German guns on the deck. I've never forgotten it.'

WO Leonard Brock was a flight engineer with No 299 Squadron in Canadian Flg Off Gib Goucher's crew:

'We took off from Keevil on "Market Garden", 17 September, but unfortu-

A formation of Stirlings flying through bursting flak shells to drop supplies on Arnhem, 19 September. (IWM BU1092)

nately our glider cast off over St Albans and we returned to base. We did, however, take part the following day with another glider and also carried out a number of re-supply drops to the troops at Arnhem. Because our normal aircraft was damaged by flak on the 19th, we flew LJ893:V on the 21st on a further supply drop but again we were hit by flak. No 7 tank was holed and we suffered damage to the wings from 88mm gunfire.'

Den Hardwick and his No 299 Squadron crew flew three sorties from Keevil to Arnhem in Stirling IV LK118. Day four of the operation (Wednesday 20th) was their thirty-second trip together as a crew.

'On day one we towed in a glider and on day three another glider carrying some of the Polish Independent Parachute Brigade, with no problems on either trip. On day four, a Wednesday, our luck ran out. Communications with the troops on the ground by this time were chaotic. We flew low over the area where they were supposed to be. Flg Off Karl Ketcheson, my Canadian bomb-aimer, was not happy and asked me to go round again as he thought he had seen them off to one side. We made another run and just as Ketchy released the containers we were hit by ground fire. Ketchy was killed immediately and the aircraft went into a dive. Ted Webb, my navigator, was in the second pilot's seat, and he and I managed to hold the dive by heaving on the two control wheels.

'Tom White, the engineer, was down at the back with two Airborne chaps who were with us to get the panniers out of the back hatch. Spotting what had happened Tom grabbed the two severed cables to the elevator trimming tabs. A pull on one cable made things worse, followed by a shout of "Wrong one!", then a pull on the other cable and we climbed away.'

Once they had gained a reasonable height Tom was able to stabilize the trimming tabs by tying the severed cables to the fuselage bulkheads with cord cut from a parachute, adjusting the settings with slip knots.

'The journey home was uneventful despite the fact that the airspeed indicator and the altimeter were not working due, we later found out, to the pitot head being damaged – a good job it was daylight. The final problem was landing. When the undercarriage and flaps are lowered an adjustment is required to the trim. We made a long approach with Tom gently adjusting the trim as required and made a safe landing, due in no small measure to the calmness of all the crew, including the two Airborne lads whom I believe were actually cooks.

'Ketchy' Ketcheson, Den Hardwick's Canadian bomb-aimer, was killed on the re-supply drop over Arnhem on 21 September. (D. Hardwick)

'On the following Saturday Bill McKee, our Canadian rear gunner from Winnipeg, was married at Bradford-on-Avon to Marguerita Fursse. The wedding was reported on the front page of the *Wiltshire News* on 29 September and the accompanying photograph, headlined "The Vacant Place", shows the crew and the bride toasting the memory of "Ketchy".'

The RAF suffered heavy losses in the ill-fated Arnhem operation between 17 and 26 September 1944, attempting to supply the beleaguered 1st Airborne Division on the ground. Forty-four Stirlings were lost and 105 aircrew killed.

CHAPTER FOURTEEN

Death over Norway

Although the fighting for Norway came to a bitter end in 1940, the struggle against the occupying German forces continued for the ensuing five years of war. Norwegian resistance movements sprang up, initially organized under the title of MILORG (MILitary ORGanization) but renamed Home Forces in the autumn of 1944. They received vital supplies of weapons and equipment from the Special Operations Executive (SOE) in England, delivered in containers by parachute from Stirlings and Halifaxes of the RAF's No 38 Group. By the war's end, a total of some 1,200 tons of equipment had been dropped to Norwegian resistance groups by the RAF's SD squadrons in 1,200 sorties.

But for the RAF's Special Duties (SD) aircrews, who flew these crucial sorties, the rugged, mountainous terrain of Norway and its unpredictable weather patterns made the locating of drop zones and the successful dropping of supplies an extremely hazardous business. Added to this was the ever-present threat of attack by German nightfighters and flak batteries. The large and heavily laden Stirlings made easy targets for the small and highly manoeuvrable German fighter aircraft like the Messerschmitt Bf 109.

As the war began to draw towards a welcome but climactic conclusion for the Allies early in 1945, it remained to be seen just how the German forces in Norway would react to capitulation. As a contingency plan, the organization and supply of weapons and equipment to Home Forces in Norway by the RAF was increased to ensure an effective striking power when the time came to confront the Germans. But the end came swiftly and with remarkably little confrontation, undermining the carefully planned strategies of the SOE and the Norwegian resistance groups. Fewer than 60,000 members of the Norwegian Home Forces, mostly carrying weapons dropped to them by RAF Stirlings and Halifaxes, successfully rounded up and disarmed 365,000 German troops in May.

The following accounts related by Norwegian historians Albert Albretsen and Jan Thygesen tell the outcomes of two such SD operations flown by Stirling crews in the opening months of 1945 to strengthen Norwegian Home Forces. Their quiet corner of southern Norway in Augst-Agder county had seen very little action during the war years, until early in 1945 three RAF transport aircraft were shot down within ten days of one another. The first of these took place in bright moonlight on 22 February.

At around 19.00hrs on 22 February 1945, No 190 Squadron's Stirling IV LK566 took off from Great Dunmow, Essex, with WO S.B. Currie at the controls, bound for Norway with a consignment of supplies for the Norwegian Home Forces. That night No 38 Group had despatched fifty-four aircraft to Norway on SD operations, although two crews aborted and in the end only thirty-four completed their vital tasks.

Flying on mission 'CROP 25', Currie's drop zone was near Gardermoen, a few miles north-east of Oslo. After making landfall in south-western Norway at Ogna Bay, west of Egersund, navigator Flt Sgt D. Hollindrake gave Currie a new course to steer and LK566 turned eastwards and followed the mountainous coastline on the landward side. In little more than 20 minutes and 100 miles further along the rocky coast the crew's luck began to run out when their lumbering Stirling, standing out clearly against the snowy landscape in the

This Browning machine-gun is being examined by Jim Marshall and his crew from No 620 Squadron. It was found in a house at Gardermoen near Oslo after the town was surrendered by the Germans in May 1945. It came from a No 190 Squadron Stirling that had been shot down by a Hauptmann Vogt, whose name and that of his wife had been painted on the side of the weapon. (J. Marshall via Stirling Aircraft Association [SAA])

bright moonlight, was spotted by a Luftwaffe nightfighter between Arendal and Tvedestrand. Albert Albretsen and Jan Thygesen take up the story:

'Both planes were soon fighting a bitter air combat and people were awakened by the sound of gunfire and the noise from the engines. In the bright moonlight they watched the German fighter make its attack. Short exchanges were fired by both aircraft but the match was unequal. The big transport aircraft flew in wide circles over the moonlit landscape fighting bravely, but it was attacked over and over again. The German was faster.

'Suddenly the Stirling caught fire. It rushed over the Lilleholt farm towards Tvedestrand fiord. People coming from a meeting watched the heavy fighting taking place above the villages of Borås and Vatnebu. Fully ablaze, the plane crashed into a rocky crag called *Lusa* – the Lice – situated in a marsh between the small communities of Langang and Torp. It skidded along the northwestern side of the crag where it exploded, much of the aircraft and its crew disappearing into the bog. At Langang people watched as the aircraft exploded and at Torp they could feel the blast. Fragments of propeller and a wing were hurled far up into the hillside towards Torp.

'Mr Knut A. Torp was one of those living close to the scene. Immediately on hearing the crash he ran down the hillside to search for survivors. He was prevented from a closer examination of the wreckage because of the danger of further explosions and the fierce heat from the fire. In any case he could not find anyone left alive. Later, when it became possible to get closer to the wreck, people were met with the terrible sight of human limbs and debris from the plane spread out over a large area.

'It did not take long before German soldiers came up from their barracks at Fiane and Tvedestrand and among the first on the scene was an Austrian ski platoon. Heavy snowfall had closed the road to Langang and Mr John Dalene, who was on his way to look for the crashed plane, was forced by the Germans to lead the way. At about the same time, more Germans arrived from Eydehavn and Arendal. They posted sentries but did not cordon the area off. The next day they began a search of the area in the hope of finding survivors, but soon concluded that nobody could have survived a crash like that.

'Between them, the German soldiers and Norwegian civilians collected what they could find of the crew's bodies – and that was not much. Much of what was left of them had already followed the Stirling into the bog. The remnants were put in a coffin and brought to Holt church nearby. After the Germans had withdrawn their guards from the crash site, quite a few people gathered at Langang in the days that followed to look for pieces from the wrecked Stirling. Many took home incredible collections of things.

'We do not know who was responsible for organizing the burial of the dead airmen but it was almost certainly due to sympathetic Norwegians. Strangely, the Germans did not oppose. The burial took place on 27 February at Holt church where people gathered from all over the district to honour and show their gratitude to the six brave airmen who had died. Never before had so many people attended a burial at Holt, so many in fact that dozens had to wait in silence outside the packed church. Inside, the coffin was placed in the chancel and covered by the Norwegian national flag. German officers were also present in the church. The church bells tolled and the vicar, Mr Danielsen, read the burial service and then gave a short sermon. The service was concluded with the Norwegian hymn *How Fair Our Country*. It was a moving ceremony. Two students from the School of Agriculture, two from the County School and two teachers carried the coffin to the cemetery. All along the route students formed a guard of honour.

'This burial was surely an extraordinary episode of the war; we have heard of no such similar episodes elsewhere in Norway.'

The crew who died in the crash were: WO S.B. Currie, 28 (pilot), Flt Sgt K.F. Newman, 21 (2nd pilot), Flt Sgt D. Hollindrake, 23 (navigator), Flt Sgt L.C. Baldock, 21 (flight engineer), Flt Sgt T.J.A. Grant, 21 (wireless operator), and Flt Sgt R.C. Davies, 22 (rear gunner).

March was a quiet month but on the night before Easter Eve, 30/31 March, SD air activity started anew with sixty-three sorties flown by No 38 Group Stirlings and Halifaxes spread across Norway, Denmark and Holland. Around midnight, three Stirlings arrived over southern Norway with supplies for the Home Forces at a DZ near Svene.

Since making landfall after their flight across the North Sea, they had already been attacked by one or more German nightfighters. If they had been flying at low level across the North Sea, the Stirlings would have avoided enemy radar contact but could have been spotted by hostile coastal shipping. Their presence would have been quickly reported to the Luftwaffe nightfighter operations centre at Grove, Denmark, with nightfighters scrambled and on their way to intercept them even before they had crossed the Norwegian coast. Albretsen and Thygesen believe the fighters came from Kjevik airfield near Kristiansand, although it is equally likely that they could have been scrambled from an airfield further south in Denmark, or even in northern Germany.

In the bright and clear moonlight conditions, the big aircraft were easily spotted by the stalking nightfighters and within minutes their fates were sealed. All three aircraft were shot down in the Augst-Agder district: No.

196 Squadron's LJ888 at Brastad near Øyestad, 3 miles west of Arendal, No 161 Squadron's LK119 at Hegland near Holt, 4½ miles west of Tvedestrand, and No 299 Squadron's LK332 at Vierli near Vegårshei, 9 miles further north.

One of these aircraft, No 161 Squadron's Stirling IV, LK119:Y, was flying mission 'BIT 14'. Y-York exploded in mid-air after it was attacked and set on fire by a German nightfighter. The remains of the aircraft and its seven-man crew, skippered by Flt Lt E.P.C. Kidd, who were on the last op of their tour, fell into thick forest at Andsmyra, Hegland near Holt. Albert Albretsen and Jan Thygesen resume their story:

> 'People in Ubergsmoen and the surrounding district were awakened in the middle of the night by the sound of a terrific battle being fought in the skies above them between a fighter and a bigger plane. Bursts of gunfire could be heard from both planes as they made big circles in the air. At one point they vanished behind the hills to the north but soon reappeared. One of the engines of the big plane caught fire but the flames soon went out. The fighter then attacked with frequent salvoes and in the end the big plane caught fire.

While flying SOE operation 'BIT 14' over southern Norway on 30 March 1945, No 161 Squadron's Stirling IV, LK119:Y, was fiercely attacked by German nightfighters causing it to explode in mid-air, killing the crew of seven captained by Flt Lt E.P.C. Kidd. The remains of the aircraft fell into thick forest at Hegland. LK119 had previously served with No 138 Squadron as NF-R, in the markings of which it appears above with Flt Lt Kidd and his crew. (J. Breeze via SAA)

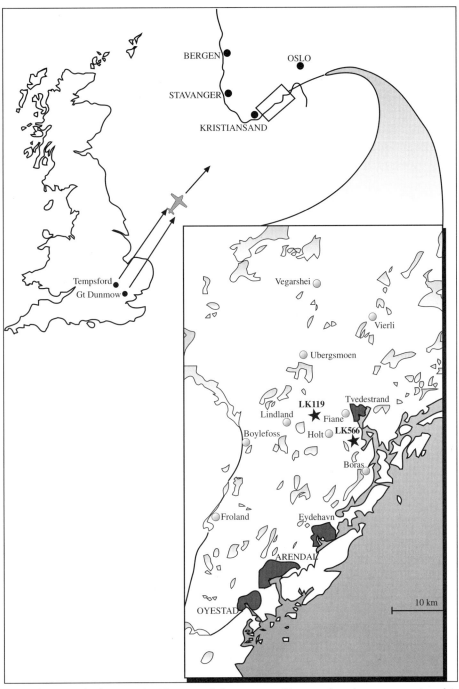

Maps showing the location in Norway of the towns, villages and settlements mentioned in connection with the crash of Flt Lt Kidd's Stirling.

The crew of Stirling LK119. *Top*: Flt Sgt Harry Minshull; *back row: left to right*, Flt Sgt R. Burgess, Flt Sgt G. Heath, Flg Off T. Macauley, Flt Lt E.P.C. Kidd; *front row*: Flt Sgt A. Shopland, WO A. Taylor. (Mrs M. Brown)

Fully ablaze it roared towards Hegland where it suddenly exploded in mid-air and crashed to the ground in the moors west of Hegland.

'At Hegland, people were awakened by the shooting and the noise. They watched how the burning torch exploded and how parts of the plane were hurled in all directions. After it had fallen to the ground, thick black smoke rose from the wreckage. Eivind Beras (now deceased) lived at Lindland nearby. Immediately following the crash he hurried to the part of the moor where he thought he'd find the plane. Draped around a cliff by the road, not far from Hegland, he found a white parachute which he hid under a rock. The snow had crusted over and the moonlight made it easy to walk through the forest.

'He reached the wreck before the Germans but found no survivors. Soldiers from the barracks at Fiane were soon on the scene and Beras tried to hide from them but he was discovered. They advanced towards him pointing their guns. They didn't know who he was so they took him into custody. After several

'X' marks the spot where Flt Lt Kidd's Stirling crashed into the forest at Hegland. In the foreground is Myklebustad Farm and the main Stavanger to Stockholm road. (Mrs M. Brown)

hours he was released and returned home in a state of shock. Later he went for the parachute but couldn't find it. Someone else had taken it.

'Tengel Hegland (also deceased) didn't live far from the scene of the crash. He, too, had seen the plane crash and hurried to the crash site where he found a parachute. As he began to gather it up the Germans arrived but, because he was a little deaf, he could not hear them shouting to him and he came close to being shot. He later went up to the wreck where he found the wings torn off, lying on each side of a small rocky cliff. The tail and forward fuselage had been broken and the four engines lay scattered in different places. The crew had been tossed out of the plane. All were dead. Germans from the barracks at Fiane arrived soon afterwards and put a guard on the wreckage.

'Four youngsters had come up from Ubergsmoen to Hegland during the night. One of them was Hakon Aslaksen. He met Sverre Furland who lived at Hegland. They found an inflatable dinghy about 6 metres long hanging in a birch tree by the road. They pulled it down and hid it up in the moors before the Germans arrived. A stirrup pump was also found along with a large number of leaflets and several parachutes, probably belonging to the containers.

Arendal Cemetery in Norway, where Flt Lt Kidd and his crew were buried. The tall obelisk in the background commemorates their sacrifice. (Mrs M. Brown)

'Over the Easter period, youngsters from Froland came up to Hegland to look for souvenirs from the crashed plane. The Stirling had been carrying a considerable load of weapons for the Home Forces and the daring boys managed to remove five rifles and a quantity of ammunition from a container, right under the noses of the German guards. One boy tried to remove the Browning machine-guns from the rear turret but in the end was forced to give up. However, the boys managed to find a jerry can amongst the wreckage and filled it with petrol from one of the plane's fuel tanks. It seems they managed to smuggle it back home and return again for a refill. But one day they were caught by the Germans who took their names and told them to disappear quickly, promising more serious treatment if they showed up again.

'Shortly after the Easter holiday the Germans arrived with a truckload of coffins. They ordered Knut Vaaje and Ole Jensen who both lived at Hegland, to bring horses and sleds. They were to drive the coffins up to the plane and down again with the bodies of the dead airmen. Per Waje and Sverre Furland from Hegland and Hakon Aslaksen and Per J. Songedal from Ubergsmoen were ordered to put the bodies into the coffins. The Germans had already collected the bodies and put them nicely together. They had found six of them close to one another, but one of the crew had lain under a spruce tree some metres away from the others. He seemed to be remarkably unhurt, but the rest of the crew had been badly mutilated.[1]

'Their thick flying clothing and boots had been removed and they had been laid out in their blue uniforms. The Germans had broken sprigs of spruce to put under the Englishmen's heads, treating the dead men with respect. While they were doing this job the Germans spotted the youngsters from Froland again and two of them were forced to put the dead airmen into their coffins. The others fled, afraid of being caught a second time.'

It was rumoured that the local priest was a Nazi sympathizer so the Norwegians arranged for a Methodist minister to officiate at the burial ceremony for Y-York's crew, at a secret location in a wood not far from the scene of the crash. Two years later the bodies were exhumed and reburied at Arendal cemetery with full military honours.

The Germans guarded the wreck for many days after the crash although eventually most of the plane was taken away and sold to scrap-metal dealers. However, some parts survived to be found again some time later – a piece of an engine cowling and a wing section bearing the serial number LK119.

Despite the many years that have passed since the wartime crash at Hegland that claimed the lives of these seven men, rumours have abounded that one of the crew members managed to escape from the doomed Stirling by parachute and flee from the Germans. Several people at Ubergsmoen claimed to have seen at least one parachute fall from the aircraft before it crashed. Two men also recalled seeing a parachutist coming down in the bog behind the home of one of them, and then disappearing towards the burning plane. However, Svein Furland, another witness living nearby, thought it might just have been the inflatable dinghy he found later hanging in a birch tree. Albretsen and Thygesen looked into the matter in some detail and this is their evaluation of the facts:

'We tried to get to the bottom of this story by talking to people in the district who were members of the wartime resistance groups, but no one knows the answer for certain. We received confirmation from the British Ministry of Defence that the crew numbered seven in total, and had the same names as those which appear on the grave stones in Arendal cemetery. The possibility that there was an eighth man – an agent – was discounted by Harald Sandvik, a former member of the prominent resistance group

Company Linge. Other former members of the group have confirmed that no other Allied airmen were in the area at that time, apart from an Australian airman named Quirk. He baled out when his plane was shot down at Holen near Arendal on 25 February.

'Some time ago, Kare Apeland from Bøylefoss recalled that just after the crash at Hegland he had found a white parachute with harness (perhaps the same one that Eivind Beras had found earlier). The 'chute was later used for making clothes but the harness still exists. When it was found there were traces of blood on it and it had been found in the same area where the parachutist was supposed to have landed.

'Others had been on their way up to the crash site that same night. Some of them tell that they had heard footsteps in the woods. When they called, the footsteps had stopped; and when they had held their calling, the steps had started again.

'Torje Vaaland arrived at the crash site early the next morning. Before he had reached the wreck he spotted a man lying face down in the moss. It looked as if he had been walking a short distance after he had reached the ground. He looked quite unhurt, with only a small blue mark at one side of his head. Could this have been the man? Did he manage to bale out and then walk the 800 metres up to the wreck where he died from exhaustion and his injuries?'

The crew of LK119 who died on the night of 30/31 March 1945 were:

Flt Lt E.P.C. Kidd, 32 (pilot), Flg Off T.S. Macaulay, 32 (bomb-aimer) Flt Sgt G.H. Heath, 22 (navigator), WO A.M. Taylor, RAAF, 24 (flight engineer), Flt Sgt H. Minshull, 20 (air gunner/despatcher), Flt Sgt A.D. Shopland, 20 (air gunner), and Sgt R.A. Burgess, 32 (wireless operator).

The RAF lost another five aircraft and crews over Norway on SD operations that night. Shepherd's Grove alone lost four of its Stirlings:

SQN	AIRCRAFT	CAPTAIN	MISSION
No 196 Sqn	Stirling IV LJ888	F/S D. Catterall	'BIT 20'
	Stirling IV LK197	P/O C. Campbell	N/K
No 298 Sqn	Halifax AVII PN243	F/O Ireland	'OSTLER 6'
No 299 Sqn	Stirling IV LK332	F/L R. Trevor-Roper	'SNAFFLE 6'
	Stirling IV PK225	F/L Anderson	'STIRRUP 8

1. According to a conversation between Mrs Mary Brown and an eyewitness to the incident, the unmutilated airman is believed to have been her brother, Flt Sgt Harry Minshull, air gunner/despatcher.

THE ROAD TO THE ORIENT

Six RAF squadrons were equipped with the Stirling V long-range freight and passenger variant between late 1944 and mid-1946. These were Nos 46, 242 (an ex-Spitfire squadron), 51, 158, 196 and 299 Squadrons. From bases in the UK they operated several flights a week to the Middle East and India, flying along routes pioneered by Imperial Airways and BOAC before and during the war.

Take, for example, a typical trooping flight flown by Stirling V, 'O', of No 158 Squadron from Stoney Cross in Hampshire between 14 and 18 September 1945. Skippered by Lt Gerald Meyer SAAF, with Plt Off Len Wilkins as co-pilot, and three crew, the route was from Stoney Cross (UK) on 14 September, to Istres (S France) to refuel, and on to Castel Benito to Lydda (Tel Aviv), with a night stop-over; 16 September, from Lydda to Shaibah (Basrah, Iraq), with a night stop-over; 17 September, from Shaibah to Mauripur (Karachi, NW India), with a night stop-over; 18 September, from Mauripur to St Thomas (Madras, SE India). The total distance flown was in the region of 6,500 miles, with a total flying time of 31hrs 40mins over five days.

Two Heavy Freight Flights (HFF) were also formed by the RAF in the autumn of 1945 to establish a regular freight service flying from the Far and Middle East back to the UK. Equipped with the Stirling V, No 1588 HFF operated five aircraft from its base at Santa Cruz (Bombay), and No 1589 HFF did the same with five more from Cairo West in Egypt.

By early in 1946, only No 46 Squadron was left flying the Stirling V on the India route, and the two Heavy Freight Flights. In July that year No 1588 HFF retired its Stirling Vs and became the last active unit in the RAF to fly the Short Stirling.

Two Heavy Freight Flights (HFF), Nos 1588 and 1589, were established in the autumn of 1945 and equipped with the Stirling V to operate a regular freight service between the Far and Middle East and the UK. No 1588 HFF's Mk V, PK178, is pictured here at Santa Cruz (Bombay) in the spring of 1946. (A. Harding via A. Thomas)

A crash on take-off from St Thomas Mount (Madras) on 13 November 1945 caused the undercarriage to collapse on No 46 Squadron's Mk V, PK173. (R. Mackay via A. Thomas)

THE ROAD TO THE ORIENT

A No 51 Squadron Mk V in a hangar at an airfield in India, after undergoing major maintenance work, 1946. (via A. Thomas)

No 51 Squadron's PK115, ORT-C, on dispersal at an Indian airfield in 1946. (J. Halley via A. Thomas)

Still in the markings of No 51 Squadron, PK148, TB-YW, awaits sale to the Belgian civil operator Trans-Air in August 1947. (E. Riding via A. Thomas)

Surrounded by the wrecks of at least two Spitfires and a Warwick, the sorry remains of PK150, ex-158 Squadron and 1589 HFF, are seen here at No 107 MU's base at Kasfareet in the Suez Canal Zone of Egypt, in 1947. The same undignified fate befell other Mk Vs – most of the HFF's Stirling Vs were broken up at Pegu airfield in Burma and sold to scrap dealers in Rangoon. (via A. Thomas)

CHAPTER FIFTEEN

The Stirling Immortalized

The Stirling's celebrity status in the British press during the early war years was largely the result of a fascination born of its great size, and from the use of the latest technology in its design and construction. Feature and news reporters were liberal in their use of superlatives when they came to describing the RAF's newest heavy bomber to enter service. Revelations about the roominess of its fuselage, warnings about the sting from its three power-operated turrets, miracle solenoid switches and the electrically retracted undercarriage, aroused great waves of interest. This fascination with the subject shines through in the enthusiastic tone of the stories they filed.

Nor was it only the national press who had something to say about this leviathan of the skies, so too did the popular and specialist magazines. *The Aeroplane*, *Flight* and *Aeronautics* magazines all published their share of illustrated articles during 1941 and 1942 extolling the virtues of the mighty and seemingly invincible Stirling, although one writer in *Flight* was driven to observe that, 'Those with an eye at all sensitive to beauty have been known to exhibit a certain lack of enthusiasm for the appearance of the Short Stirling . . . [but] nobody can possibly dislike the look of a Stirling more than do the Germans!'. This fact was indisputably confirmed in 'Stirling Squadron', a bullish four-page illustrated article which appeared in *Flight* on 23 October 1941, describing a visit to No 15 Squadron at Wyton where crews reportedly considered that in any scrap with a German fighter they would be successful 'nine times out of ten'. Sadly, Bomber Command's loss statistics would ultimately prove to be the converse.

An eyewitness impression of the manufacture and assembly of Stirlings graced the pages of the March 1942 issue of *Aeronautics*, while an article entitled, 'I am an aircraft designer' in *The Aeroplane* of 19 June 1942, featured the transcript of a BBC radio interview with Arthur Gouge, who had directed the design of the Stirling at Shorts during the late 1930s. Later that year, on 11 September, 'A Stirling Job' looked at what the Stirling was like to fly from the viewpoint of Capt F.D. Braybrooke, an Air Transport Auxiliary pilot. Not to be outdone by

Charles E. Brown photographed these Stirling Is of No 1651 HCU for *Picture Post*, in flight over the Cambridgeshire countryside on 30 April 1942. (IWM CH5475)

the specialist aviation press, the month of September also saw Hulton's national weekly magazine *Picture Post* present 'A Stirling Carries Eight Tons of Bombs', a feature relating the significant contribution of the RAF's four-engined heavies to the war effort. Accompanying the article were the superb air-to-air photographs by the renowned photographer Charles E. Brown, taken on 30 April 1942 and featuring the Stirlings of No 1651 Heavy Conversion Unit at Waterbeach and one of its pilot instructors, Flt Sgt George Mackie. The year 1943 witnessed the 25th anniversary of the formation of the RAF. *Everybody's Weekly* of 3 April 1943 marked the occasion by featuring a striking view of a Stirling on its cover as a backdrop to a pilot in full flying kit.

Despite the dark purpose for which the Stirling had been designed, the cartoonists of *Punch* saw through to the rich vein of Stirling humour that was just waiting to be tapped. Almost without exception, its great size was satirized in the sharply observed drawings of David Langdon, Rowland Emett and Russell Brockbank, whose work graced the pages of the magazine through the war years.

When it came to the big screen, however, unlike its cousins the Vickers Wellington and the Avro Lancaster, the Stirling was noticeable by its absence. As the backbone of the RAF's early bomber offensive, the Wimpy gained international renown in the war years from its starring roles in the popular films *Target For Tonight* (Dir. Harry Watt, 1941, featuring aircraft of No 149 Squadron) and *One of Our Aircraft is Missing* (Dir. Michael Powell & Emeric Pressburger,

Flt Sgt George Mackie at the controls of a Stirling I at Waterbeach, 30 April 1942. (Author's collection)

1942). It comes as a surprise to learn that the clamour of publicity surrounding the arrival of the RAF's first four-engined heavy bomber was never translated into a film for the big screen. It is conceivable that films featuring either Nos 7 or 15 Squadrons, both of which were equipped with the Stirling when *Target for Tonight* was being shot, would have seemed the obvious choice for the Ministry of Information (MoI) who commissioned it. Indeed, the Stirling's only feature film appearance became a brief 30-second bit-part by No 7 Squadron at the very end of *One of Our Aircraft is Missing*.

However, a 30-minute documentary film commissioned by the MoI, entitled *Speed Up on Stirlings*, was released in about 1942 and tells how the early Stirlings were built, and of the men and women who built them. The different phases of work in progress can be seen underway at the Shorts Shadow Factory at South Marston, near Swindon in Wiltshire.

Postwar, the Avro Lancaster came to epitomize the wartime exploits of RAF Bomber Command and soon became the centrepiece of a number of movies celebrating the fact – *Appointment in London* (Dir. Philip Leacock, 1953) and *The Dam Busters* (Dir. Michael Anderson, 1954). But even if the postwar filmmakers had wanted to feature the Stirling, by then this aircraft type was obsolete

'OKAY – COME ON! LEFT HAND WELL DOWN.' *Punch* cartoonist David Langdon's impression of the giant Stirling, 1944. (Reproduced by permission of *Punch*)

and perhaps – more pertinently – there were none left to film, having already fallen to the scrap-men's blow torches.

The question of why no popular film featuring the Stirling was ever made becomes all the more surprising when one considers the part played by the novelist H.E. Bates in the Air Ministry's Public Relations Department, PR.11. He was recruited by the RAF in 1941 as a short story writer, and given the substantive rank of flight lieutenant. He was granted considerable freedom of movement and access to the Stirling crews of No 7 Squadron at Oakington where he arrived in November 1941. His task was to experience the atmosphere of an operational RAF bomber station from first-hand and then to write short stories about the life he observed. The Air Ministry envisaged his stay at Oakington as being no more than a few days but Bates, realizing the intensity of his task, assessed that it would more likely become two or three weeks, perhaps even two or three months.

Under the pseudonym Flying Officer X, he wrote a number of short stories which appeared almost immediately in the *News Chronicle* during 1942. In the second volume of his autobiography, *The World in Ripeness*, H.E. Bates recalls:

'The stories duly appeared in *The News Chronicle* under that *nom de plume*, which in fact hid my identity for just about a single day. My imprint, it seemed, was on every word. When I got back to Oakington the first of the *News Chronicle* stories had already been widely read and I felt a sense of inexplicable unease. There was much leg-pulling; and the Wing Commander, I was told, was gunning for me. The Station Commander was fortunately far more tolerant . . . Later it was of no little satisfaction to me to know that the men of Bomber Command themselves approved of the stories . . .'[1]

With the backing of the head of Air Ministry Public Relations, Hilary St George Saunders, the Flying Officer X stories were collected and published in book form by Jonathan Cape as *The Greatest People in the World* (1942) and *How Sleep the Brave* (1943). Keenly observed from the viewpoints of the Stirling aircrew with whom he had mingled, these cameos of war in the air are among the finest ever written. The longest story, entitled *How Sleep the Brave,* tells of a Stirling crew whose aircraft is badly shot up and set on fire after bombing the target, and who are eventually forced to ditch in the North Sea. After sixty hours adrift in the sea, by a quirk of sheer good fortune their dinghy is washed up on the shores of England.

When Bates was informed that the first printing of his Flying Officer X stories would be 100,000, his reaction was, 'Never having heard of a book, except the Bible and *Gone With the Wind*, selling such figures, I nearly fainted.'[2]

The end of the war saw many former servicemen and women with stirring tales to tell put pen to paper. One of the many excellent real-life adventure stories to make it into print was *Boldness Be My Friend* by Richard Pape, the true account of a No 15 Squadron Stirling navigator shot down over Holland in September 1941 after raiding Berlin. His aircraft became one of the first Stirlings to fall reasonably intact into enemy hands. First published by Elek Books in 1953, *Boldness Be My Friend* very quickly became a bestseller and sold some half-a-million copies in paperback alone. Pape recounts the astonishing adventures of how four times he escaped his captors before finally gaining his freedom, exactly three years to the day after being shot down, by faking the symptoms of a fatal disease.

The exigencies of war thrust together many unlikely partners. This was very much the case with the art world which came into contact with the Stirling, its builders and its crews, in a number of different ways. Dozens of artists and designers were retained by the MoI during the war years to paint and draw a wide range of subjects for possible future use by the Ministry as publicity material – for instance, a 100ft high canvas mural of a Stirling was used to promote Birmingham's 'Wings for Victory' week in July 1943. One artist in particular among the great fund of talent was the painter Laura Knight RA (later Dame Laura, 1877–1970), who was to be acclaimed as one of the greatest British painters of the twentieth century. She was asked to design a poster as a warning against careless working practices that had caused fires, destroying many factories engaged in important war work. In her autobiography she tells how,

Dame Laura Knight's 'Take Off, Interior of a Bomber Aircraft', 72in x 60in, oil, 1944. (IWM Cat No. LD 3834)

'This poster, which I had offered to do without pay, was placed in every workshop in England and happily turned out to be such a success that I received a very handsome cheque for it.'[3] While preparing studies for this poster in the Austin Shadow Factory at Longbridge she conceived the idea of making a water-colour study of the construction of Stirlings. 'It was impossible to paint in oil because of the confusion of different lighting. It had to be a big picture . . . A paper-hanger employed on war work joined my sheets of Whatman's "handmade" together for me and pasted them down on an enormous board. On this I painted, supported by an easel of tubular construction.'[4] The huge painting that resulted was bought by the managing director of Austin, Sir Leonard Lord, and exhibited at the Royal Academy's exhibition in the summer of 1944. It can be seen today at the British Motor Industry Heritage Trust, Gaydon, Warks.

Notable among her other works is a 6ft x 5ft oil painting entitled 'Take Off, Interior of a Bomber Aircraft'. It was completed in 1944 and was very finely observed; it represents the scene in a Stirling cockpit on take-off showing the pilot, second pilot, navigator and flight engineer busy at their respective duties. She reportedly chose a Stirling of No 15 Squadron at Mildenhall because the type was soon to be phased out by the squadron and replaced by the Lancaster. An idle Stirling was put at Laura Knight's disposal for her to study and make preparatory sketches before she made her final painting. Later she painted a large pierrot on the nose of this aircraft beneath the pilot's window, with the legend 'Midgley's Flying Circus' above. This almost certainly refers to Flt Lt D. Midgley DFC who was a pilot on the squadron at this time. The model for the navigator in this painting was in fact a wireless operator with No 15 Squadron named Ray Escreet. As a flying officer, DFM, he was on his first tour of ops in late 1943 when asked to model by the artist. Sadly, he was later to die on an abortive SOE mission in March 1945 when his Hudson aircraft was shot down over Luxembourg by 'friendly' fire. Later, when Laura Knight heard of his death, she was concerned that his mother should receive a photograph of the painting.

It was clear from the ways in which the Stirling was interpreted and perceived by journalists, writers, painters and film-makers during the war years, that its contribution to the war effort went far beyond the obvious one of carrying the air war to the German homeland. It is all the more surprising therefore, to observe that since the war's end only three books have been published about the Stirling, with a handful more in which it is mentioned in passing.

1. H.E. Bates, *The World in Ripeness* (Michael Joseph, 1972), p. 21
2. Op cit, p. 30
3. Laura Knight DBE, RA, *The Magic of a Line* (William Kimber, 1965)
4. Op cit

IN MEMORIAM

In common with all RAF aircrew during the Second World War, the men who flew in Stirlings were volunteers. Just like their contemporaries who flew Spitfires, Lancasters, Beaufighters or Sunderlands, once they had volunteered for aircrew and had been accepted, they were committed to finishing their tour of operations. For better or worse they were in, but could not volunteer out if things were subsequently not to their liking. For the vast majority of Bomber Command aircrew there were only two ways of leaving operational flying: the first was to complete a tour of thirty ops; the second was to die on operations.

Some Stirling aircrew achieved recognition in terms of bravery awards for their contribution to ridding the world of Naziism, while others were content simply to have played their part and survived long enough to enjoy the fruits of peace. But for every man who lived to tell the tale, there were many more aircrew – numbering 55,000 from Bomber Command alone – who gave their lives flying on operations. Some have graves in the foreign countries where they fell, but many more vanished without trace and consequently have no known grave.

The photographs of Stirling aircrew that follow represent some of the silent many who, in that expressive official euphemism of the war years, simply 'failed to return'. For each face that appears below, the likenesses of many more who left the shores of England never to return have simply been forgotten. Lest we forget.

IN MEMORIAM

A mute but powerful reminder of what the Commonwealth Air Forces Memorial at Runnymede in Surrey stands for. (Author's collection)

The Gates of Remembrance: Runnymede, where the names of RAF and Commonwealth Air Forces aircrew who have no known grave are commemorated. (Author's collection)

PLT OFF RALPH LONSDALE RNZAF, PILOT, No 149 SQUADRON, AGE 21.
Returning from ops to Aachen on 6 October 1942 with an engine on fire, Ralph Lonsdale ordered his crew to bale out. He stayed with his Stirling to the end and perished when it crashed near Faversham, Kent. (J. Brigden)

FLG OFF TED RATCLIFFE RAFVR, NAVIGATOR, No 15 SQUADRON, AGE 30.
On the homeward journey from a raid on Wilhelmshaven on 19 February 1943, Ted Ratcliffe's Stirling was attacked and shot down into the North Sea by a Messerschmitt Bf 110 nightfighter. His body was eventually recovered and buried on the Dutch island of Ameland. (C. Gilbert)

IN MEMORIAM

SGT ALBERT EDWARDS RAFVR, BOMB-AIMER, No 90 SQUADRON, AGE 31.
Albert Edwards and his entire crew went missing on 12/13 May 1943, target Duisberg. It was their tenth op together. They have no known graves and their names are recorded on the Allied Air Forces Memorial at Runnymede. (Mrs S. Ennis)

SGT GEORGE JACQUES RAFVR, BOMB-AIMER, No 218 SQUADRON, AGE 20.
George Jacques' Stirling was shot down by a nightfighter at Lichtenvoorde, Holland, at 01.17hrs on 26 June 1943, returning from a raid on Gelsenkirchen. From the crew of seven only the navigator survived the ordeal.
(P. Jacques)

SGT GILBERT ROSE RAFVR, FLIGHT ENGINEER, No 218 SQUADRON, AGE 32.
Attacked by an enemy aircraft returning from a raid on Mulheim in the early hours of 23 June 1943, Gilbert Rose's Stirling later crashed into the North Sea off the Dutch coast. Only the rear gunner survived to become a PoW. The bodies of Gilbert Rose and two other crew members were never found. (Miss Y. Reynolds)

SGT DENNIS 'LOFTY' MATTHEWS RAF, FLIGHT ENGINEER, No 196 SQUADRON, AGE NOT KNOWN.
On a re-supply drop to Arnhem on 21 September 1944, 'Lofty' Matthews' Stirling was badly shot up over the drop zone where he and the rest of the crew baled out, only to be shot to pieces in their harnesses as they floated to earth. The sole survivor was the pilot.

Appendix I

SHORT STIRLING I – LEADING PARTICULARS

Name: Stirling I
Type: Four-engined midwing landplane
Duty: Heavy bomber

MAIN DIMENSIONS
Span: 99ft 1.12in
Length overall (tail down): 87ft 1.7in
Height (tail down): 22ft 9in

MAIN PLANE
Aerofoil section: Gottingen 436 (modified)
Chord (theoretical) at fuselage centreline: 21.888ft
Dihedral (top surface at front and rear spar booms): 2° 25'
Sweepback (leading edge to fuselage datum): 5° 0'

TAILPLANE AND ELEVATOR
Aerofoil section: RAF 30 (modified)
Span: 40ft 8.5in
Chord at root: 9ft 2.5in
Dihedral: nil

FIN AND RUDDER
Aerofoil section: RAF 30 (modified)
Chord at root: 9ft 11.12in

FUSELAGE
Length overall (rigging position): 87ft 3.5in
Height (over fin stub without coupé and aerial mast): 9ft 10.75in
Width maximum: 6ft 7.5in

AREAS

Main planes, total (with ailerons & flaps): 1,322sq ft
   Ailerons, total: 117.6sq ft
   Flaps, total: 405sq ft
Tail planes, total (with elevators): 239sq ft
   Elevators, total (with tabs); 93sq ft
   Elevator trimming tabs, total: 2.96sq ft
Fin (with rudder): 77.68sq ft
   Rudder (with tabs): 33.58sq ft
   Rudder balance tab: 0.725sq ft
   Rudder trimming tab: 0.845sq ft

UNDERCARRIAGE

1. MAIN WHEEL UNITS
Type: Single-wheel, twin shock-strut units retracting into engine nacelles
Track: 25ft 7.85in

Shock-struts: 'Short' and 'Turner' designs
Type: Pneumatic-hydraulic

Wheels
Type: Dunlop AH2233
Tyre: Dunlop IBB14 (26.5in x 21in)
Tyre pressure: 42lb/sq in

Brakes
Type: Dunlop pneumatic
Pressures (max in container): 300lb/sq in
      (min effective): 100lb/sq in

2. TAIL WHEEL UNIT
Type: Two single-wheel, single shock-strut units retracting into fuselage
Track: 23.5in

Shock-struts: 'Dowty' design
Type: Pneumatic-hydraulic

Wheels
Type: Dunlop wheelless aero
Tyre: Dunlop WS14 (10in x 5.25in)
Tyre pressure: 50lb/sq in

LEADING PARTICULARS

ENGINES
Name: Hercules II or XI
Type: 14–cylinder, air-cooled, supercharged, two-row radial

AIRSCREWS
Type: de Havilland 3–bladed variable pitch, metal, with constant speed control
Diameter: 13ft 6in

TANK CAPACITIES

Fuel

Leading edge	Tank No 7 two, each	179gal
Inboard	Tank No 2 two, each	331gal
Inboard middle	Tank No 4 two, each	254gal
Outboard middle	Tank No 5 two, each	164gal
Outboard	Tank No 6 two, each	81gal
Inner rear	Tank No 1 two, each	80gal
Outer rear	Tank No 3 two, each	63gal

Max capacity: 2,304gal

Oil

Oil space, four tanks, each	25.5gal
Air space, four tanks, each	5.5gal

Total oil capacity: 102gal

Appendix II

PRINCIPAL STIRLING MANUFACTURING SITES

SHORT BROTHERS LTD, ROCHESTER, KENT
Seaplane Works, Rochester – Main factory
Strood Extension – Shadow factory
Cuxton – Design, drawing and stress office
Rochester Airport – Final assembly, flight-testing

SHORT & HARLAND LTD, BELFAST, NORTHERN IRELAND
Queen's Island, Belfast – Main factory
Sydenham Airport – Final assembly, test-flying, delivery
Aldergrove, Long Kesh, Maghaberry – Final assembly, flight-testing, delivery

AUSTIN MOTOR CO. LTD, LONGBRIDGE, BIRMINGHAM
Austin Shadow Factory, Longbridge – Main factory
Austin Airframe Shadow Factory, Cofton Hackett – Components
Rover Aero Factory No 1, Acock's Green – Components
Rover Aero Factory No 2, Solihull – Components
Marston Green Shadow Factory – Final assembly
Elmdon Airport – Flight-testing, delivery

SHORT BROTHERS LTD – Temporary Dispersed Production Sites, SWINDON, WILTS
South Marston – Shadow Factory
Blunsdon – Shadow Factory
Highworth – Shadow Factory
Stratton St Margaret – Drawing office, production control

OTHER TEMPORARY DISPERSED SITES
Kidderminster, Worcs. – Design, drawing and stress office
Hucclecote, Glos. – Shadow Factory

PRINCIPAL MANUFACTURING SITES

SHORT BROTHERS REPAIR ORGANIZATION (SEBRO), CAMBRIDGE

Madingley Road, Cambridge – Fuselage/component repair and salvage
Bourn Airfield, Cambridge – Mainplane repair, re-assembly and test-flighting

SUBCONTRACTORS

These are just some of the many hundreds of dispersed subcontractors who produced components for the Stirling:

Bristol Aeroplane Co. Filton, Bristol	engines, instrument panels
Carrs Paper Mill, Shirley, Birmingham	bomb-doors, horizontal stabilizers
Lea Francis Engineering Ltd, Coventry	fuel tank lids, throttle boxes, mudguards, oxygen bottle containers
Mackie's Ltd, Belfast	tailplane, control column, main undercarriage components
Ultra Electric Ltd, Acton, London	bomb-doors, elevators, rudders instrument panels, blind approach equipment
S. Graham Ross Ltd, Slough	ancillary equipment
Ferodo Ltd, Chapel-en-le-Frith	brake linings
Reynolds Tube Co. Ltd, Birmingham	aluminium sheet, extrusions etc
James Booth & Co. Ltd, Birmingham	aluminium sheet, extrusions etc
Delco-Remy & Hyatt Ltd, London	electric motors
Rubery Owen & Co. Ltd, Birmingham & Coventry	aircraft pressings, sheet metal fabricating etc
Exactor Control Co. Ltd, London	hydraulic remote controls for engines
Fisher & Ludlow Ltd, Birmingham	aircraft pressings, sheet metal fabricating etc
Sperry Gyroscope Co. Ltd, Brentford	navigational instruments
BRICO Ltd, Coventry	piston rings
Richard Klinger Ltd, Sidcup, Kent	jointing, gaskets for exhaust and cylinder heads
Dunlop Ltd, Birmingham & Belfast	tyres, wheels
De Havilland Propeller Division	propellers
Smiths Ltd	instruments

Appendix III

PRINCIPAL STIRLING AIRFIELDS AND OPERATORS 1940–6

Dates refer to periods that squadrons/units were equipped with the Stirling. In several instances squadrons/units were already in residence at a particular station before, and for a period after, these dates, operating different aircraft types. An asterisk denotes a minor user of the Stirling, in some cases as few as one aircraft.

The number preceding the entry for each homebased airfield refers to its location on one of the accompanying maps. (NB: Longtown, Nutts Corner, Talbenny and Tilstock do not appear on the maps.)

1. ALCONBURY, Cambs. (satellite to WYTON)

2. BLYTON, Lincs.
   *1662 HCU

3. BOTTESFORD, Lincs.
   90 Sqn   07/11–29/12/42

4. BOURN, Cambs. (satellite to OAKINGTON)
   7 Sqn   06/41–07/42
   15 Sqn   13/08/42–14/04/43

5. CHEDBURGH, Suffolk
   214 Sqn   01/10/42–10/12/43
   620 Sqn   17/06–22/11/43

6. DOWNHAM MARKET, Norfolk
   214 Sqn   10/12/43–06/01/44
   218 Sqn   08/07/42–07/03/44
   623 Sqn   10/08–06/12/43

AIRFIELDS AND OPERATORS 1940–6

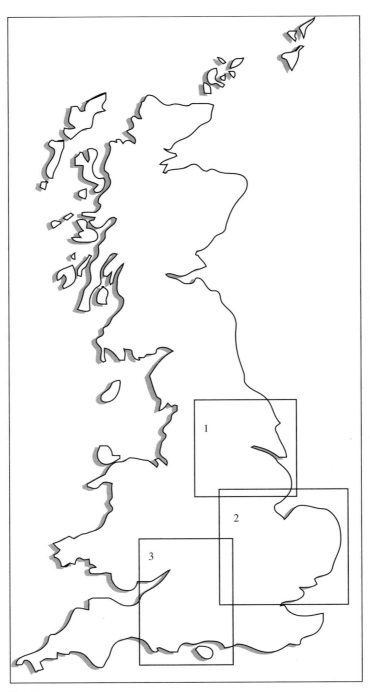

Key Map

7. FAIRFORD, Glos.
   190 Sqn   25/03–14/10/44
   620 Sqn   18/03–18/10/44

8. GREAT DUNMOW, Essex
   190 Sqn   14/10/44–06/45
   620 Sqn   18/10/44–07/45

9. HARWELL, Oxon
   295 Sqn   15/03–11/10/44
   570 Sqn   14/03–08/10/44

10. HULLAVINGTON, Wilts.
    ★1427 Flt   05–08/42

11. KEEVIL, Wilts.
    196 Sqn   14/03–09/10/44
    299 Sqn   15/03–09/10/44

12. LAKENHEATH, Suffolk
    149 Sqn (Det)   11/41–04/42
    199 Sqn   05/07/43–01/05/44

13. LECONFIELD, Yorks.
    51 Sqn   01/06–21/08/45

14. LEEMING, Yorks.
    7 Sqn   01/08–29/10/40

15. LEICESTER EAST, Leics.
    190 Sqn   05/01–25/03/44
    196 Sqn   18/11/43–07/01/44
    620 Sqn   22/11/43–18/03/44

16. LISSETT, Yorks.
    158 Sqn   29/05–17/08/45

    LONGTOWN, Cumberland
    1332 HTCU   05/09–07/10/44

AIRFIELDS AND OPERATORS 1940-6

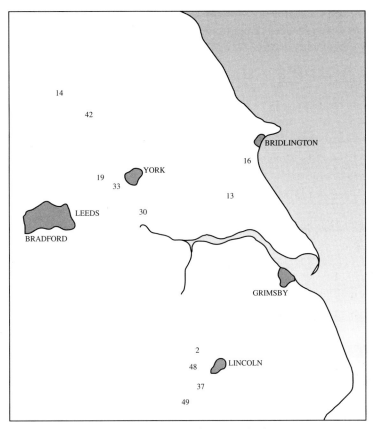

Location Map 1: Stirling airfields in north-east England.

17. LYNEHAM, Wilts.
    ★525 Sqn   29/05–27/07/44

18. MARHAM, Norfolk
    218 Sqn   16/12/41–08/07/42
    ★1427 Flt 08/42–10/43

19. MARSTON MOOR, North Yorks.
    ★1652 HCU

20. MATCHING, Essex
    O & RTU   03–10/45

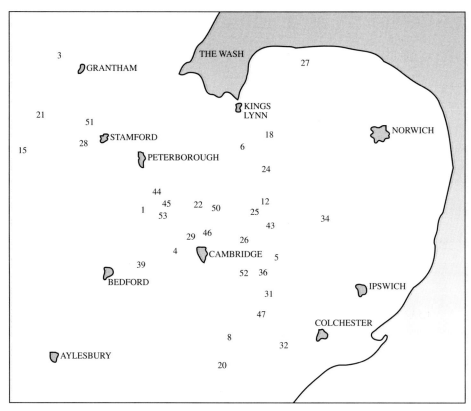

Location Map 2: Stirling airfields in the East Midlands and East Anglia.

21. MELTON MOWBRAY, Leics.
    1588 HFF   16/09–10/45
    1589 HFF   28/09–10/10/45
    ★304 FTU/12 FTU 1944–5

22. MEPAL, Cambs.
    75 Sqn   28/06/43–28/04/44

23. MERRYFIELD, Somerset
    242 Sqn   09/12/45–01/05/46

24. METHWOLD, Norfolk
    149 Sqn   15/05–18/09/44
    218 Sqn   04–17/08/44

AIRFIELDS AND OPERATORS 1940-6

25. MILDENHALL, Suffolk
    15 Sqn   14/04–27/12/43
    149 Sqn   02/11/41–06/04/42
    622 Sqn   10/08/43–09/01/44

26. NEWMARKET, Suffolk
    7 Sqn (Det)   27/03–27/04/41
    75 Sqn   01/11/42–28/06/43

27. NORTH CREAKE, Norfolk
    *171 Sqn   08/09/44–31/03/45
    199 Sqn   01/05/44–14/03/45

28. NORTH LUFFENHAM, Rutland
    1653 HCU 12/44

    NUTTS CORNER, Co. Antrim, Northern Ireland
    1332 HTCU   07/10/44–25/04/45

29. OAKINGTON, Cambs.
    7 Sqn   29/10/40–29/08/43
    1657 HCU 10/42

30. RICCALL, North Yorks.
    1332 HTCU   04/45

31. RIDGEWELL, Essex
    90 Sqn   29/12/42–31/05/43

32. RIVENHALL, Essex
    295 Sqn   11/10/44–15/01/46
    570 Sqn   08/10/44–08/01/46

33. RUFFORTH, North Yorks.
    *1663 HCU

34. SHEPHERD'S GROVE, Essex
    196 Sqn   26/01/45–27/03/46
    299 Sqn   25/01/45–15/02/46

35. STONEY CROSS, Hants.
    46 Sqn   07/01/45–12/04/46
    242 Sqn   15/02/45–01/05/46
    299 Sqn   04/11/43–15/03/44

36. STRADISHALL, Suffolk
    51 Sqn   21/08/45–25/07/46
    158 Sqn   17/08/45–01/01/46
    214 Sqn   12/01–01/10/42
    1657 HCU   10/42–12/44
    ★1427 Flt   02/10/43 (incorporated into 1657 HCU)

37. SWINDERBY, Lincs.
    1660 HCU   11/43–02/45

    TALBENNY, Dyfed
    ★303 FTU/HFU 1944–5

38. TARRANT RUSHTON, Dorset
    196 Sqn   07/01–14/04/44
    295 Sqn   12/45–01/46

39. TEMPSFORD, Beds.
    138 Sqn   11/06/44–10/03/45
    161 Sqn   05/09/44–02/05/45

40. THRUXTON, Wilts.
    ★1427 Flt   12/41–05/42

41. TILSTOCK, Shropshire
    1665 HCU   23/01/44–03/46

42. TOPCLIFFE, Yorks.
    ★1659 HCU

43. TUDDENHAM, Suffolk
    90 Sqn   13/10/43–07/06/44

44. UPWOOD, Hunts.
    ★PFF NTU

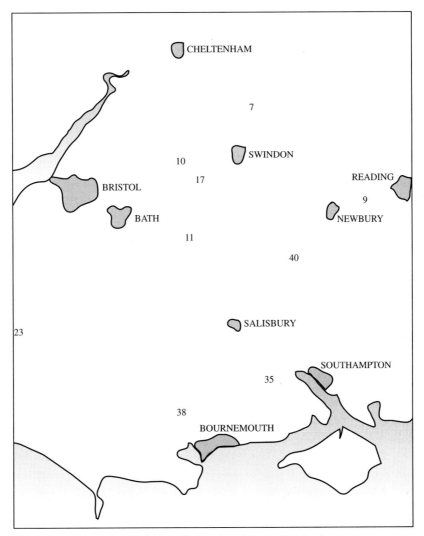

Location Map 3: Stirling airfields in the south and west of England

45. WARBOYS, Cambs. (satellite to WYTON)
    *PFF NTU

46. WATERBEACH, Cambs.
    1651 HCU 02/01/42–21/11/43
    1665 HCU 01/05–05/06/43

47. WETHERSFIELD, Essex
   196 Sqn   09/10/44–16/03/46
   299 Sqn   09/10/44–15/02/46
   ★O&RTU

48. WIGSLEY, Notts.
   ★1654 HCU

49. WINTHORPE, Notts.
   1661 HCU 09/11/43–28/02/45

50. WITCHFORD, Cambs.
   513 Sqn   15/09–24/12/43

51. WOOLFOX LODGE, Rutland
   1651 HCU 10/11/44–05/01/45
   1665 HCU 05/06/43–23/01/44

52. WRATTING COMMON (West Wickham), Cambs.
   90 Sqn   31/05–13/10/43
   1651 HCU 21/11/43–10/11/44

53. WYTON, Cambs.
   15 Sqn   14/03/41–13/08/42

OVERSEAS BASES

BLIDA, Algeria
624 Sqn   07/06–24/09/44

BRINDISI, Italy
148 Sqn   11–12/44

CAIRO WEST, Egypt
1589 HFF   10/45–04/46

DRIGH ROAD (Karachi), India
1588 HFF   10/45

SANTA CRUZ (Bombay), India
1588 HFF   13/10/45–20/05/46

Appendix IV

STIRLING SQUADRONS AND THEIR COMMANDING OFFICERS – 1940–6

Following each squadron number is the mark(s) of Stirling flown and the period during which the squadron operated the type; and from which airfield(s) followed by the month in which the squadron first took up residency there; following the name of each officer commanding is his date of appointment.

No 7 SQUADRON
(Mk I, III – 08/40–08/43)
Leeming, Yorks.   08/40
Oakington, Cambs.   10/40

No 7 Squadron, A Flight groundcrew with Wg Cdr Hamish Mahaddie, May 1943. (H. Mahaddie)

Wg Cdr P. Harris	08/40
Wg Cdr H.R. Graham	04/41
Wg Cdr B.D. Sellick DFC & Bar	04/42
Wg Cdr O.R. Donaldson DSO, DFC	10/42
Wg Cdr H.H. Burnell	05/43

No 3 Group – 918 sorties, 41 aircraft lost (4.5%)
No 8 (PFF) Group – 826 sorties, 37 aircraft lost (4.5%)
27 Stirlings destroyed in crashes
Introduced the Stirling into RAF service
An original Pathfinder Force (PFF) squadron
Suffered the highest percentage losses of Stirling squadrons
Last PFF Stirling sortie on 10 August 1943

No 15 SQUADRON
(Mk I, III – 04/41–12/43)

Wyton, Hunts.	05/40
Bourn, Cambs.	08/42
Mildenhall, Suffolk	04/43

Wg Cdr H.R. Dale	04/41 (POW 11/05/41)
Wg Cdr P.B.B. Ogilvie DSO	05/41
Wg Cdr S. Menaul DFC, AFC	12/41
Wg Cdr D.J.H. Lay DSO	*c.* 10/42?
Wg Cdr J.D. Stephens	05/43

No 3 Group – 2,231 sorties, 91 aircraft lost (4.1%)
38 Stirlings destroyed in crashes
Flew more bombing raids than any other Stirling squadron
Suffered the heaviest losses (with 218 Sqn) of all Stirling squadrons

No 46 (UGANDA) SQUADRON
(Mk IV, V – 02/45–05/46)

Stoney Cross, Hants.	01/45

Wg Cdr B.A. Coventry	01/45
Wg Cdr S.G. Baggott	12/45
Wg Cdr R.G. Dutton DSO, DFC	03/46

No 75 (New Zealand) Squadron, B Flight aircrew, December 1943. (75 Sqn Association)

No 51 SQUADRON
(Mk IV, V – 06/45–04/46)
Leconfield, Yorks.	04/45
Stradishall, Suffolk	08/45

Wg Cdr E.F.E. Barnard	04/45

No 75 (NEW ZEALAND) SQUADRON
(Mk I, III – 11/42–04/44)
Newmarket, Suffolk	11/42
Mepal, Cambs.	06/43

Wg Cdr V. Mitchell	11/42 (KIA 18/12/42)
Sqn Ldr G.T. Fowler	12/42
Wg Cdr G.A. Lane	01/43

Wg Cdr M. Wyatt   05/43
Wg Cdr R.D. Max DFC   08/43

No 3 Group – 1,736 sorties, 72 aircraft lost (4.1%)
The only New Zealand night bomber squadron in Bomber Command

No 90 SQUADRON
(Mk I, III – 11/42–06/44)
Bottesford, Leics.   11/42
Ridgewell, Essex   12/42
West Wickham/Wratting Common, Cambs.   05/43
Tuddenham, Suffolk   10/43

Wg Cdr J.C. Clayton DFC   12/42
Wg Cdr J.H. Giles DFC   06/43
Wg Cdr G.T. Wynne-Powell DFC   12/43
Wg Cdr F.M. Milligan   01/44
Wg Cdr A.J. Ogilvie   06/44

No 3 Group – 1,937 sorties, 58 aircraft lost (3.0%)
26 Stirlings destroyed in crashes

No 138 (SD) SQUADRON
(Mk IV – 06/44–03/45)
Tempsford, Beds.   03/42

Wg Cdr W.J. Burnett DFC   05/44
Wg Cdr T.B.C. Murray   12/44

No 3 Group – 503 sorties, 10 aircraft lost (1.9%)
Carried out operations in support of Resistance groups

No 148 (SD) SQUADRON
(Mk IV – 11–12/44)
Brindisi, Italy   01/44

Wg Cdr D.C. Hayward DSO   08/44

Flew four Stirlings alongside Halifax and Lysander types in support of Resistance groups

No 149 Squadron aircrew, June 1942. Flt Sgt Middleton VC is in the back row, *second from left*. (Mrs N. Curtis)

No 149 (EAST INDIA) SQUADRON
(Mk I, III – 11/41–09/44)

Mildenhall, Suffolk	04/37
Lakenheath, Suffolk	04/42
Methwold, Norfolk	05/44

Wg Cdr G.J. Spence	11/41
Wg Cdr C. Charlton-Jones	05/42 (KIA 29/08/42)
Wg Cdr K.M. Wasse DFC	09/42
Wg Cdr G.E. Harrison DFC	04/43
Wg Cdr C.R.B. Wigfall	09/43
Wg Cdr M.E. Pickford	05/44

No 3 Group – 2,628 sorties, 87 aircraft lost (3.3%)
40 Stirlings destroyed in crashes
Flew the most Stirling sorties in Bomber Command
Posthumous VC awarded to Flt Sgt R.H. Middleton, Turin, 28–29/11/42

No 158 SQUADRON
(Mk IV, V – 05–12/45)
Lisset, Yorks.	02/45
Stradishall, Suffolk	08/45
Wg Cdr F.J. Austin DFC	05/45
Wg Cdr D. Iveson DSO, DFC	07/45
Wg Cdr P. Dobson	12/45

No 161 (SD) SQUADRON
(Mk IV – 09/44–06/45)
Tempsford, Beds.	04/42
Wg Cdr L.M. Hodges DSO, DFC	05/43
Wg Cdr G. Watson DFM	01/45
Wg Cdr M.A. Brogan DFC	02/45
Wg Cdr L.F. Ratcliff DSO, DFC, AFC	03/45

No 3 Group – 379 sorties, 6 aircraft lost (1.6%)
Carried out operations in support of Resistance groups

No 171 (RCM) SQUADRON
(Mk III – 09/44–01/45)
North Creake, Norfolk	09/44
Wg Cdr M.W. Renaut DFC	09/44

(Sqn flew one flight of Stirling IIIs alongside Halifax IIIs)

No 100 Group – 171 sorties, no losses

No 190 SQUADRON
(Mk IV – 01/44–06/45)
Leicester East, Leics.	01/44
Fairford, Glos.	03/44
Great Dunmow, Essex	10/44
Wg Cdr G.E. Harrison DFC	01/44 (KIA 21/09/44)
Wg Cdr R.H. Bunker DSO, DFC & Bar	10/44
Wg Cdr G.H. Briggs DFC	04/45

No 196 Squadron, Bombing Section, late 1944. (J. Hibbs)

No 196 SQUADRON
(Mk III, IV, V – 07/43–03/46)
Witchford, Cambs.	07/43
Leicester East, Leics.	11/43
Tarrant Rushton, Dorset	01/44
Keevil, Wilts.	03/44
Wethersfield, Essex	10/44
Shepherd's Grove, Suffolk	01/45

Wg Cdr N. Alexander		12/43
Wg Cdr M.W.L. Baker	08/44 (KIA 21/02/45)	
Wg Cdr T.F. Turner DFC, MC		02/45
Wg Cdr J. Blackburn DSO, DFC		01/46

No 3 Group – 166 sorties, 11 aircraft lost (6.6%)
No 38 Group – N/K
Transferred from Bomber Command to No 38 Group in November 1943 for airborne forces and transport duties

No 214 Squadron groundcrews, Stradishall, mid-1942. (J. Hardman)

SQUADRONS & COMMANDING OFFICERS

Canadian aircrew serving with Nos 149 and 199 Squadrons at Lakenheath, April 1944. (199 Register)

No 199 (RCM) SQUADRON
(Mk III – 07/43–03/45)

Lakenheath, Suffolk	06/43
North Creake, Norfolk	05/44
Sqn Ldr G.T. Wynne-Powell	02/43
Wg Cdr L.M. Howard	04/43
Wg Cdr N.A.N. Bray DFC	10/43
Sqn Ldr Lumsdain	?/44
Sqn Ldr W.A. Betts	08/44
Wg Cdr Bennington	?/45

No 3 Group – 681 sorties, 14 aircraft lost (2.1%)
No 100 Group – 1,378 sorties, 4 aircraft lost (0.3%)
Transferred to No 100 Group in May 1944 as an RCM squadron

No 214 (FEDERATED MALAY STATES) SQUADRON
(Mk I, III – 04/42–02/44)

Stradishall, Suffolk	01/42
Chedburgh, Suffolk	10/42
Downham Market, Norfolk	12/43

No 218 Squadron aircrew, Downham Market, September 1943. (J. McIlhinney)

Wg Cdr A.H. Smythe	09/42
Wg Cdr M.V.M. Clube	03/43
Wg Cdr D.J. McGlinn DFC	07/43
Wg Cdr D.D. Rogers	08/44

No 3 Group – 1,432 sorties, 54 aircraft lost (3.8%)
29 Stirlings destroyed in crashes

No 218 (GOLD COAST) SQUADRON
(Mk I, III – 01/42–08/44)

Marham, Norfolk	11/40
Downham Market, Norfolk	07/42
Woolfox Lodge, Rutland	03/44

Wg Cdr P. Holder DSO, DFC	01/42
Wg Cdr O.A. Morris	08/42
Wg Cdr D. Saville DSO, DFC	03/43 (KIA 25/07/43)
Wg Cdr W.G. Oldbury OBE, DFC	08/43
Wg Cdr D.D. Rogers	03/44

No 570 Squadron air and groundcrews, Rivenhall, November 1945. (N. Harrison)

No 3 Group – 2,600 sorties, 91 aircraft lost (3.5%)
35 Stirlings destroyed in crashes
Posthumous VC awarded to Flt Sgt A. Aaron DFM, Turin, 12–13/08/43
Suffered the heaviest losses (with 15 Sqn) of the Stirling squadrons

No 242 (CANADIAN) SQUADRON
(Mk IV – 02/45–01/46)
Stoney Cross, Hants.   11/44

Wg Cdr H. Burton DSO, MBE   12/44?
Wg Cdr D.W. Balden   01/45
Wg Cdr D.M. Walbourn DSO   02/45
Wg Cdr E.J. Wicht DSO, DFC   12/45

No 295 SQUADRON
(Mk IV – 06/44–01/46)
Harwell, Oxon   03/44
Rivenhall, Essex   10/44

Wg Cdr B.R. Macnamara   05/43
Wg Cdr H.E. Angell   09/44
Wg Cdr R.N. Stidolph   11/45

No 299 SQUADRON
(Mk IV, V – 01/44–02/46)
Stoney Cross, Hants.   11/43

Keevil, Wilts.   03/44
Wethersfield, Essex   10/44
Shepherd's Grove, Suffolk   01/45

Wg Cdr P.B.N. Davis DSO   12/43 (KIA 19/09/44)
Wg Cdr C.B.R. Colenso DFC   09/44
Wg Cdr P.N. Jennings   12/44
Wg Cdr R.N. Stidolph   09/45

No 513 SQUADRON
(Mk III – 10–11/43)
Witchford, Cambs.   09/43

Wg Cdr G.E. Harrison DFC   09/43

No 570 SQUADRON
(Mk IV – 07/44–12/45)
Harwell, Oxon   03/44
Rivenhall, Essex   10/44

Wg Cdr R.J.M. Bangay   11/43
Wg Cdr K.R. Slater   06/45
Wg Cdr R.E. Young DSO, DFC   08/45
Wg Cdr J. Blackburn DSO, DFC   12/45

No 620 SQUADRON
(Mk I, III, IV – 06/43–07/45)
Chedburgh, Suffolk   06/43
Leicester East, Leics.   11/43
Fairford, Glos.   03/44
Great Dunmow, Essex   10/44

Wg Cdr D.H. Lee DFC   06/43 (MIA 22/09/44)
Wg Cdr G.T. Wynne-Powell DFC   10/44

No 3 Group – 339 sorties, 17 aircraft lost (5.0%)
9 Stirlings destroyed in crashes
No 38 Group – N/K
Transferred from Bomber Command to No 38 Group in November 1943 for airborne forces and transport duties

No 622 SQUADRON
(Mk III – 08/43–01/44)
Mildenhall, Suffolk   08/43

Sqn Ldr J. Martin   10–20/08/43 (acting)
Wg Cdr G.H.N. Gibson DFC, AFC   08/43–01/44
No 3 Group – 195 sorties, 7 aircraft lost (1.7%)
2 Stirlings destroyed in crashes

No 623 SQUADRON
(Mk III – 08–12/43)
Downham Market, Norfolk   08/43

Wg Cdr E.J. Little DFC   08/43
Wg Cdr G.T. Wynne-Powell   09/43
Wg Cdr F.M. Milligan AFC   11/43

No 3 Group – 150 sorties, 10 aircraft lost (6.7%)
1 Stirling destroyed in crash

No 624 (SD) SQUADRON
(Mk IV – 07–09/44)
Blida, Italy   02/44

Wg Cdr C.S.G. Stanbury DSO, DFC   01/44

(Sortie and attrition data: *The Bomber Command War Diaries*, Part 3: Operational Performances of Units)

Appendix V

NAMED AND PERSONALIZED STIRLINGS

The desire of air and groundcrews to personalize their individual aircraft with painted names, badges, emblems, or images, or a combination of all four, was a practice more common to the US Army Air Force than to the RAF during the Second World War. Officially, King's Regulations forbade such adornment of Air Force property, but many enlightened squadron commanders – particularly in Bomber Command – allowed their crews to indulge in a little artistic or calligraphic enterprise.

Bomber aircraft, by virtue of their sheer size, provided a greater surface area than a small fighter over which to trail lettering or emblazon some fancy artwork. The designs and lettering drew upon a variety of influences – the most common were the symbols denoting the number of ops flown by a particular aircraft. These varied from the simple bomb – usually one colour for night ops and another for daylights (rarely the latter for Stirlings), to ice-cream cones representing trips to Italy, daggers (SOE ops), gliders (airborne ops) and parachutes (paratroop drops).

Favourite subjects were invariably inspired by the female form derived from the name of the pilot's wife or from the aircraft's call-sign; the name of a sponsoring country, group of people or an individual (eg, 'MacRobert's Reply'); titles of popular songs or names of cartoon characters, and even brands of cigarettes were also used as inspiration. The permutations were as numerous as the aircraft they adorned. The artwork itself was of widely varying standards and depended very much upon who could be found to apply the paint to the aircraft. Some pieces were quite literally works of art and produced to a very high standard; at Mildenhall a pierrot was painted on to a No 15 Squadron Stirling by Laura Knight (later Dame Laura, RA), the famous painter.

The following list is by no means an exhaustive one, but goes some way towards illustrating the wide range of names applied to the slab sides of the RAF's Stirlings. Information is recorded in the following order: name, mark of aircraft, serial number, squadron(s), date in use.

Sugar was an unidentified Mk I which probably belonged to No 7 Squadron during 1941–2. (Author's collection)

SUGAR – Mk I, No 7 Sqn/15 Sqn?, mid-1941

THE JOKER – SEMPER IN EXCRETA ('always in the shit'), No 149 Sqn/620 Sqn? 02/43 (LK386:J?)

EAST INDIA 1 – Mk I, OJ-E, N6103, No 149 Sqn, 11/41–05/42

EAST INDIA 2 – Mk I, OJ-Q, N6122, No 149 Sqn, 12/41–06/42

EAST INDIA 2/3? – Mk I, OJ-G, R9295, No 149 Sqn, ?–03/42

MacROBERT'S REPLY – Mk I, LS-F, N6086, No 15 Sqn, 09/41–02/42

MacROBERT'S REPLY – Mk I, LS-F, W7531, No 15 Sqn FTR 02–05/42 (FTR)

THE LAST DIP IN GOLDEN FLEECE – Mk III, LK619, 1332 HCU, 1944–5

Photographed at Rivenhall in April 1945 with more than thirty smiling groundcrew arrayed before her, No 570 Squadron's Mk IV, LK292, W-Witch, shows off her fuselage artwork. Flying on a broomstick while towing a Horsa glider behind her, the witch drops canisters of supplies to

the Resistance. Operational symbols denote two bombing raids, one SD operation and a German fighter shot down. (B. Gibbs)

The Latin motto on the side of this unidentified Stirling of No 149 Squadron, in the spring of 1943, reads *Semper in Excreta* which roughly translated means 'Always in the Shit'. Also of interest are the twenty-six ops that are recorded in the form of eight-pointed stars, and two nightfighter kills (possibly a Bf 110 and a Ju 88). (B. Gibbs)

555 State Express was Mk IV, LK555, of No 570 Squadron stationed at Rivenhall, Essex, in the spring of 1945. She was the personal 'mount' of Sqn Ldr J. Stewart DFC, B Flight commander. (J. Stewart DFC)

NAMED AND PERSONALIZED STIRLINGS

Pictured here during the summer of 1944, Jolly Roger was Mk III, LJ525, which served with No 199 (SD) Squadron at North Creake, Norfolk. She is pictured here with her air and groundcrews. (R. Smith)

THE NUTHOUSE – Mk III, OJ-N, EE963, No 149 Sqn, 08/43
IT'S IN THE BAG – Mk IV, ZO/7T-H/Z, ?, No 196 Sqn, 1944
THE GREMLIN TEASER – Mk III, EX-G, LJ542, No 199 Sqn, 04–12/44
YORKSHIRE ROSE II – Mk IV, QS-L & D4-Y, LJ566, No 620 Sqn, 05/44–06/45
CHEERS FOR BEER AT THE GETSUMINN – Mk IV, QS-?, ?, No 620 Sqn, 1944
GOOFY II – Mk IV, 8E-O, EF446, No 295 Sqn, 08/44–07/45
THE BUSHWACKER – Mk IV, 8Z-H, LJ995, No 295 Sqn, 08/44–02/45
SHORTY GEORGE – Mk IV, 8Z-Y?, ?, No 295 Sqn, summer 1944
BEER IS BEST – Mk IV, V8-B, LJ977, No 570 Sqn, 10–12/44
SHOOTING STARS – Mk IV, 'WES', LK171, Rivenhall C/O's personal a/c, 08–11/44
555 STATE EXPRESS – Mk IV, E7-S, LK555, No 570 Sqn, 07/44–10/45
KISMET III – Mk IV, 8Z/8E-?, ?, No 295 Sqn, 1944

No 295 Squadron's Mk IV, Thunderbird III, carries an image of a Canadian Indian thunderbird Totem on its side along with a variety of different symbols to denote SD operations (daggers), bombing sorties (bombs), paratroop drops (parachute), and glider operations (gliders). (via Chaz Bowyer)

THE SAINT – Mk IV, 5G-?, EF267, No 299 Sqn, summer 1944
BARNEY'S BULL – Mk III, EX-?, ?, No 199 Sqn, 1943
JOLLY ROGER – Mk III, EX-R, LJ525, No 199 Sqn, 04/44–01/45
GLORIOUS BEER – Mk IV, 8Z-B, LK129, No 295 Sqn, 08/44–04/45
EN AVANT – Mk III, EX-E, LJ513, No 199 Sqn, 04–11/44
JUST JAKE – Mk I/III?, BU-?, ?, No 214 Sqn, ?
ROTHWELL'S RUFFIANS II – Mk III, HA-D, BK803, No 218 Sqn, 06–06/43
TE KOOTI – Mk III, LS-U, BK611, No 15 Sqn, 24/12/42–FTR 05/43
THUNDERBIRD III – Mk IV, 8Z/8E-?, ?, No 295 Sqn, summer 1944
CRUSADER – Mk I Srs III, BU-C, R9152, No 214 Sqn, summer 1942
L'S A COMIN' – Mk IV, E7-L, ?, No 570 Sqn, autumn 1944
SITTING PRETTY – Mk IV, MA-?, ?, No 161 Sqn, 1944
BITTER SWEET – Mk IV, 5G-B, EF322, No 299 Sqn, summer 1944
MIDGLEY'S FLYING CIRCUS – Mk III, LS-?, ?, 15 Sqn, autumn 1943

Bibliography

1. PRIMARY SOURCES:

a. PUBLIC RECORD OFFICE, KEW, SURREY

AIR 27 203	No 15 Squadron ORB (1/41–12/43)
AIR 27 494	No 51 Squadron ORB (1–12/45)
AIR 27 646–7	No 75 Squadron ORB (1/42–12/43; 1–12/44)
AIR 27 731–2	No 90 Squadron ORB (3/37–12/43; 1–12/44)
AIR 27 956	No 138 Squadron ORB (8/41–12/44)
AIR 27 1002–4	No 149 Squadron ORB (1–12/42; 1–12/43; 1/44–5/45)
AIR 27 1050	No 158 Squadron ORB (10/44–12/45)
AIR 27 1172	No 199 Squadron ORB (11/42–7/45)
AIR 27 1321–3	No 214 Squadron ORB (1–12/42; 1–12/43; 1–12/44)
AIR 27 1350	No 218 Squadron ORB (1–12/42)
AIR 27 1472	No 242 Squadron ORB (1/43–12/45)
AIR 27 1654	No 299 Squadron ORB (11/43–1/46)

b. PRIVATE SOURCES

Interviews and correspondence with former aircrews and groundcrews, whose names are listed in the acknowledgements section at the front of this book. Transcript of the interview of WO Edgley, 15 Squadron, Bomber Command, RAF, by MI9/IS9(W) on 29 August 1945.

2. SECONDARY SOURCES:

a. BOOKS

Alwyn Philips, J. *Valley of the Shadow of Death: The Bomber Command Campaign March–July 1943* (Air Research Publications, 1992)

Bates, H.E. *The World in Ripeness* (Michael Joseph, 1972)

Bomber Command Continues: The Air Ministry Account of the Rising Offensive Against Germany, July 1941–June 1942 (HMSO, 1942)

Bowyer, Michael J.F. *The Stirling Bomber* (Faber, 1980)

Cooper, Alan W. *In Action With The Enemy* (William Kimber, 1986)
Durand, Arthur A. *Stalag Luft III: The Secret Story* (Louisiana UP, 1988)
Falconer, Jonathan *Stirling at War* (Ian Allan Ltd, 1991)
Ford–Jones, Martyn R. *Bomber Squadron: Men Who Flew With XV* (William Kimber, 1987)
Gomersall, Bryce *The Stirling File* (Air–Britain Ltd, 1979)
Green, William & Swanborough, Gordon *World War 2 Aircraft Fact Files: RAF Bombers Part 2* (Jane's, 1981)
Jefford, Wg Cdr C.G. *RAF Squadrons* (Airlife, 1988)
Knight, Dame Laura *The Magic of a Line* (William Kimber, 1965)
Mackay, Ron *Short Stirling in Action* (Squadron/Signal Publications, 1989)
Merrick, Ken *Flights of the Forgotten: Special Duties Operations in World War 2* (Arms & Armour Press, 1989)
Middlebrook, Martin *Arnhem 1944: The Airborne Battle* (Viking Penguin Books Ltd, 1994)
Middlebrook Martin & Everett, Chris *The Bomber Command War Diaries* (Viking Penguin Books Ltd, 1985)
Neave, Airey *Saturday at M.I.9: A History of Underground Escape Lines in North West Europe in 1940–5* (Hodder and Stoughton, 1969)
Rawlings, John D.R. *Coastal, Support and Special Squadrons of the RAF and Their Aircraft* (Jane's, 1982)
Searby, Air Cdre John *The Everlasting Arms* ed. Martin Middlebrook (William Kimber, 1988)
Woodley, Charles *Golden Age: Commercial Aviation in Britain 1945–1965* (Airlife, 1992)

b. NEWSPAPERS, PERIODICALS AND MAGAZINES:

The Times, FlyPast, Aeronautics, Sunday Pictorial, Flight, The Aeroplane, Picture Post

Index

(Page numbers in parentheses indicate footnotes)

Aaron, Sgt A. 86
Abwehr 65
Adams, Gp Capt 20
Aeronautical Inspection
    Directorate 50
Aeronautics 161
Air Ministry 67, 165
Air Transport Auxiliary 2, 29,
    161
    No 8 Ferry Pilots Pool 29
A&AEE xv, 1
Albretsen, A. 146, 148, 149, 150,
    155
Algeria 86
Alps 37, 38, 39, 41, 42, 86, 87,
    120
Anderson, Flt Lt 156
Anderson, M. 163
Apeland, K. 156
Appointment in London 163
Arnott, Sgt P. 54, 56, 57
Ashbaugh, WO 90
Ashton, Flg Off 99
Aslaksen, H. 153, 154
Atkins, Sgt J. 86, 87, 92
Atlantic Ocean 20
Austin 'Bunny' 8
Austin, Sgt 1
Austin Motor Co., Longbridge
    23, 24, 26, 28, 30, 44,
    67, 112, 167
Australia
    Sydney 54

Bailey, B. 107
Bailey, Sgt J. 93, 94, 96, 99
Bailey, Flt Sgt M. 21
Baker, F. 129, 131
Baldock, Flt Sgt L. 149
Barker, Flt Sgt J. 137
Bates, H.E. 165
Battle of Berlin 1, 93
Battle of North Cape 21
Battle of the Ruhr 54, 108
Bay of Biscay 20
Beaton, Sgt L. 86, 87, 89

Belfast 68
Belgium 61, 62, 65, 91
    Aachen 170; Achel 63;
    Antwerp 63; Ardennes forest
    91; Brussels 63, 64; Lommel
    63; Mons 64; Neerpelt 63;
    Overpelt 63
Bennett, Sgt B. (99)
Bennett, AVM D. 99
Beras, E. 152, 156
Berridge, Flt Lt G. 40
Berry, Flt Lt F. 94, 99
Bettles, Flg Off A. 121
Bickerson, Flg Off B. 39, 40,
    41
Birmingham 67, 70, 71, 72, 166
    Broad Street 67, 72; Carr's
    Paper Mill, Shirley 30; Elmdon
    67, 70; Swansborough Park,
    King's Heath 31
Blacklock, Flg Off G. 1, 8–11,
    14, 16, 20
Blackpool Illuminations 46
Blackwell, J. 113
Blenheim, Bristol 8
B-17 Flying Fortress (xiv), 101
B-24 Liberator (xiv)
Boggis, Flg Off P. 1, 16
Boldness be my Friend 166
Bomber Command Continues 23
Boorman, R. 5
Borthwick, A. 113, 114
Boyce, Flg Off F. 135
Bradford-on-Avon, Wilts. 145
Braybrooke, Capt F. 161
Bristol Aeroplane Co. 32
British European Airways 2
BOAC 157
Broadstairs, Kent 91
Brock, WO L. 128, 131, 142
Brockbank, R. 162
Brown, B. 46
Brown, Charles E. 162
Bulkington, Wilts. 67
Bull, C. 49, 52
Burgess, Flt Sgt R. 152, 156

Burns, Sgt B. 108, 109, 110, 112,
    117, 118
Burroughs, Flg Off 99

Cambridge 8, 43, 45, 47, 53, 116
    Coton 45; King's College 43;
    Madingley Road 43, 44;
    Sidney Street 53
Campbell, Plt Off C. 156
Canada
    British Columbia 54;
        Winnipeg 145
'Captain', the 64, 65
Carless, D. 49
Catterall, Flt Sgt D. 156
Channel Islands 20
Chappell, Flt Lt B. 119, 120, 122,
    123
Chappell, E. 10
Chelmsford, Essex 124
Chile
    Concepción 54
Chippenham Lodge, Suffolk 112
Circuit, J. 52, 53
Clark, Flg Off J. 135
Cleaver, Mrs 63
Cole, Sgt B. 64
Collins, Flt Lt 16, 17
Collins, Sqn Ldr 1, 2
Commonwealth Air Forces
    Memorial 169, 171
Cooper, Plt Off B. 54, 56, 57
Copeman, Flt Lt G. 131, 134,
    135
Cornelius, J. 63
Cosgrove, Sgt H. 86, 87, 89
Coventry 31
Cox, Plt Off R. 1, 2, 8, 20
Craven, Sgt 64
Cripps, Sir S. 49
Cruickshank, Flt Lt 8
Currie, WO S. 147, 149
Curtis, Plt Off S. 125, 128
Cuxton, Kent 5

Dalene, J. 148

Danielsen, Mr 149
Davies, A. 30
Davies, Flt Sgt R. 149
Dawe, Flt Lt G. 29
D-Day 108, 123, 124, 129–35
de Minchin, R. 133
Denmark 90, 98, 125, 149
  Frisian Islands 16, 90; Grove 149; Langelands Belt 42; Taagerup 40
Dill, Gen Sir J. 2
Donaldson, Sgt H. (99)
Drake, Flt Eng W. 29
Durrans, Flt Sgt E. (99), 112–16

Eastbourne, Sussex 115, 116
Edgley, Sgt A. 54, 56, 59, 60, 61, 65
Edwards, Sgt A. 171
Egypt
  Cairo West 157; Kasfareet 160
Emett, R. 71, 162
English Channel 41, 42, 114, 124, 131, 134
Entwhistle, Plt Off 118
Escreet, R. 167
Everybody's Weekly 162

Fahy, Flt Sgt L. 124
Faulconbridge, T. 113, 114, 115
Faversham, Kent 170
Field, WO D. 108, 112, 115, 116
Flamborough Head, Yorks. (21)
Flight 22, 161
Flying Officer X 165
Foxton, H. 122
France 61, 65, 87, 109, 120, 124, 128, 130, 134
  Aulnoye 108; Blois 112; Bordeaux 64; Bourges 112; Brest 8, 14, 20, 21; Brittany 20; Caen canal 134; Chartres 89; Cherbourg 89; Dieppe 42; Fresnes prison 65; Istres 157; La Pallice 20, 21; La Rochelle 20; Le Havre 131; Le Mans 112; Loire, river 112; Montdidier 118; Orne, river 130, 131; Paris 38, 39, 42, 64, 86, 87; Strasbourg 123, 124
Fuersse, M. 145
Furland, S. 153, 154, 155

Garretts, Flt Sgt B. 137
Gedge, B. 46
Gedge, G. 49
Germany 66, 71, 74, 101, 105
  Berlin xiii, 11, 14, 54, 87, 93, 94, 96, 97, 101, 166; Borkum 21; Bremen 14, 101; Bremerhaven 16; Buckeburg 84; Cochem 91; Cologne 11; Dresden 65; Duisberg 171; Dulag Luft, Oberursel 65; Düsseldorf 54, 56, 59; Elbe, river 40; Emden 11, 14, 16; Essen (xiv); Fallersleben 42; Flensburg 90; Frankfurt 65, 101; Gelsenkirchen 102, 171; Hamburg 108; Hanover 14, 101, 105, 106; Heligoland 40, 90; Kassel 101; Kiel 14; Lingen 11, (21); Ludwigshafen 101; Magdeburg 14; Mainz 91; Mannheim 101; Muhlberg am Elbe 65; Mulheim 172; München Gladbach 103; Nuremberg 94; Peenemünde 108; Rhine, river 123, 134, 136; Ruhr valley 54, 71, 102; Stuttgart 42; Wilhelmshaven 170
Gestapo 65
Gill, Plt Off J. 86, 89
Gilman, Mrs 49
Glass, Sqn Ldr R. 38, 133
Godfrey, T. 98
Goodenough, Sgt 64
Goucher, Flg Off G. 128, 131, 142
Gouge, A. 5, 161
Graham, 'Jock' 8, 10, 15, 16, 17, 20
Graham, Wg Cdr R. 14
Grant, Flt Sgt T. 149
Griffiths-Jones, Sqn Ldr J. 8

Halifax, Handley Page xiii, (xiv), 21, 23, 38, 85, 86, 87, 100, 101, 116, 146, 149, 156
Halkett, Flt Sgt A. 66
Hall, Sgt F. 39, 42
Hampden, Handley Page xiii
Hardwick, Plt Off D. 123, 124, 125, 130, 132, 144, 145
Harris, ACM Sir A. 99, 128
Harris, Sqn Ldr 1, 2
Harris, T. 32
Harvie, I. 121
Haywards E. 49, 52
Heath, Flt Sgt G. 152, 156
Hegland, T. 153
Heinkel He 177 (xiv)
Hercules, Bristol (engine) 25, 28, 37, 55, 74–6, 122
Hipper 8

Hiscock, Sgt S. 98
Holland 61, 83, 136, 140, 149, 166, 171
  Ameland 170; Arnhem 134, 136, 138, 140–5, 172; Budel 62, 63; Eindhoven 62; Grubbenvorst 56, 59; Helmond 61; Lichtenvoorde 171; Meijel 62; Rotterdam 8, 10; Venlo 59, 61; Weert 62
Holland, Capt G. 130, 131, 132
Hollindrake, Flt Sgt D. 147, 149
Holt, Norfolk 54
Hooker, Flt Sgt J. 137
Horsa, Airspeed 129, 131, 133, 134, 135, 136, 140, 141
How Fair Our Country 149
How Sleep The Brave 165
Howard-Smith, Sqn Ldr G. 8, 10
Hoysted, Flg Off H. 125, 133, 136, 137, 140, 142
Hudson, Lockheed 167
Hughes, Flt Capt J. 29
Hugo, Sgt F. 64
Hynam, Sgt A. 94, 95, 99

Icklingham, Suffolk 109
Ilford, Essex 54
Imperial Airways 2, 157
India 157, 158
  Mauripur 157; Pegu (Burma) 160; Rangoon 160; Santa Cruz (Bombay) 158; St Thomas Mount 157, 158
Iraq
  Shaibah 157
Ireland, Flg Off 156
IS9 59
Italy 86, 87
  Genoa 37, 39; La Spezia 37; Milan 37; Turin 37, 38, 86, 94

Jacques, Sgt G. 171
Jensen, O. 154
Jerman, Sgt J. 90
Jerromes, Sgt. C. 42
Johnen, W. 85
Jones, Sgt O. (99)
Junkers Ju 88 xiii, 94, 95, 97, 101, 108, 109, 125

Ketcheson, Flt Sgt K. 124, 144, 145
Kidd, Flt Lt E. 150, 151, 152, 153, 154, 156
King's Lynn, Norfolk 39
Knight, Dame L. 166, 167

Lancaster, Avro xiii, (xiv), 23, 38,

INDEX

71, 86, 87, 100, 101, 108, 117, 162, 163, 168
Langdon, D. 162, 164
Larden, Flt Sgt A. (99)
Lea Francis Engineering 31
Leacock, P. 163
Leake, B. 46
Leigh-Mallory, AM Sir T. 67
Letchworth Training Centre 46, 49, 51
Levett, Sgt M. 89
Libya
  Castel Benito 157
Lloyd, G. 71
London 32, 63, 66, 67, 70, 116
  Edmonton 54; St Paul's Cathedral 70; Trafalgar Square 66
Long Kesh, NI xviii
Lonsdale, Plt Off R. 170
Lord, Sir L. 167
'Louise' 64
Luftwaffe 58, 85, 148, 149, 150
  Units: 1./NJG1 85; 7./NJG1 (21); I./JG52 (21)
Luxembourg 167
Lynch-Blosse, Sqn Ldr P. 8, 14

Macauley, Flt Off T. 152, 156
Mackie, Flt Sgt G. 162, 163
Mackie, G. 121, 122
Maes, C. 64
Mahaddie, H. 38
Marsh, Flt Sgt G. 93–9
Marsh, LAC L. 74, 78
Marshall, Flt Lt J. 147
Marshall, Flt Lt K. 84
Marshall's of Cambridge 53
Martin, Lt 135
Mason, Sgt C. 86, 89, 91
Matthews, Sgt D. 142, 172
May, Sqn Ldr R. 42
Maxted, Sgt S. 54, 56, 57, 59, 61, 63, 64, 65
McCleod, Flg Sgt G. 124
McKee, B. 145
Meaburn, Sgt J. 94
Mediterranean Sea 86
Melksham, Wilts. 131
Messenger, Capt 2
Messerschmitt
  Bf 109, 15, 16, 17, 20, (21), 146; Bf 110, 11, 85, 86, 91, 170; Me 410, 19
Metcalfe, J. 140, 142
Meyer, Lt G. 157
Midgley, Flt Lt D. 167
Miller, Glen 53

Milligan, A. 110, 117
MI9 65
Ministry of
  Aircraft Production 30, 31, 32; Defence 155; Information 6, 22, 163, 166; Labour 44, 46, 49
Minshull, Flt Sgt H. 152, 156
Mitchell, Plt Off D. 119, 120, 122
Moloney, Sgt L. 64
Molyneaux, Sgt 1
Mont Blanc 37
Morris, Sgt J. 39, 40
Morris, Wg Cdr O. 42

Naish, A. 14
National Savings 67
Nesbit, B. 46
Nesbit, F. 49
Newman, Flt Sgt K. 149
News Chronicle 165
Newton Flotman, Norfolk 14
North Sea xiii, 16, (21), 54, 98, 125, 149, 165, 170, 172
Norway 146–51, 154
  Arendal 148, 150, 154, 155, 156; Augst-Agder county 146, 149; Boras 148; Boylefoss 156; Brastad 150; Egersund 147; Eydehavn 148; Fiane 148, 152, 153; Froland 154, 155; Gardermoen 147; Hegland 150, 152, 153, 154, 155, 156; Holen 156; Holt 148, 149, 150; Kjevik 149; Kristiansand 149; Langang 148; Lilleholt 148; Lindeland 152; Myklebustad 153; Oslo 147; Oyestad 150; Stavanger 153; Svene 149; Torp 148; Tvedestrand 148, 150; Ubergsmoen 150, 153, 154, 155; Vatnebu 148; Vegarshei 150; Vierli 150

Old, A. 5
Old Dorothy Café Ballroom Band 53
One of our Aircraft is Missing 162
Owen, Flt Sgt R. 137

Pacey, I. 44–5
Palestine
  Lydda 157
Pape, R. 166
Parker, J. 5
Parkinson, Flt Sgt G. 42
Pathfinder Force *(see Royal Air Force, 8 Group)*

Peden, Flg Off M. 100, 101, 105, 106
Petlyakov Pe-8 (xiv)
Piaggio P108B (xiv)
Picture Post 22, 162
Pike, Flt Lt V. 14
Pittard, Sgt R. 54, 57
Portsmouth, Hants. 90
Powell, J. 110, 112
Powell, M. 162
Pressburger, E. 162
Price, 'Taffy' 8, 16, 17, 20
Proctor, Percival 29
Prowd, WO K. 133, 142
Prune, B. 66
Punch 71, 162, 164
Pyrenees 124

Queen's Island, Belfast 29

Ratcliffe, Flg Off T. 170
Rattigan, M. 31
Read, Flt Lt 125
Richards Sgt L. 42
Richards, Plt Off R. 94, 96, 99
Richardson, Sgt 135
Roberts, J. 136
Rose, Sgt G. 172
Rosen, E. 32
Rossett, C. 63
Rossiter (WOp) 8, 11, 20
Rowley Mile, Suffolk 11
Royal Academy 167
Royal Air Force
  Bomber Command xiv, 2, 9, 22, 29, 37, 54, 71, 85, 93, 101, 108, 119, 161, 163, 168
  Groups
  3 Group 10, 21, 29, 43, 54, 86, 108; 4 Group 21; 8 Group 86, 94
  38 Group 119, 129, 136, 146
  Squadrons
  7 Squadron 2, 8, 10, 11, 12, 14, 16, 20, 21, 37, 38, 42, 64, 70, 74, 79, 163, 165; 15 Squadron 15, 16, 37, 42, 54, 60, 66, 68, 70, 93, (99), 103, 161, 163, 166, 167, 170; 35 Squadron 21; 46 Squadron 157, 158; 51 Squadron 157, 159, 160; 70 Squadron 92; 75 Squadron (99), 118; 76 Squadron 21; 90 Squadron 42, (99), 108, 109, 110, 112, 117, 171; 99 Squadron 1; 106 Squadron 38; 138 Squadron 120, 125; 149 Squadron 1, 8,

17, 18, 42, 86, 88, 90, 91, 92, 119, 123, 124, 162, 170; 158 Squadron 157, 160; 161 Squadron 120, 150; 190 Squadron 147; 196 Squadron 125, 129, 130, 133, 136, 137, 149, 156, 157, 172; 199 Squadron 84, (99); 214 Squadron 38, 42, 100, 101, 105; 218 Squadron 37, 39, 42, 64, 86, (99), 171, 172; 242 Squadron 157; 295 Squadron 83; 299 Squadron 123, 128, 129, 130, 142, 144, 156, 157; 570 Squadron 84; 620 Squadron 138, 147; 622 Squadron 93, (99); 623 Squadron (99); 1588 HFF 157, 158; 1589 HFF 157, 158, 160; 1651 HCU 86, 162; 1657 HCU 18; 1665 HCU 133; 11 OTU 86; ;107 MU 160; B Per T Flt 1, 2

Airfields
Alconbury xv; Boscombe Down xv, 1, 8; Bourn 43, 44, 46, 47; Carpiquet (B17), France 125; Chedburgh 80, 101; Downham Market 39, 40, 41; Fairford 138; Farnborough 83; Finningley 2; Friston 112, 115, 116; Great Dunmow 147; Halton 116; Keevil 123, 125, 128, 129, 131, 133, 134, 136, 142, 144; Lakenheath 86, 87, 92, 123, 124; Lee-on-Solent 136; Ludford Magna 125; Manston 41, 84; Methwold 119; Mildenhall 17, 18, 54, 55, 58, 66, 98, 99, 108, 167; Newmarket 11, 74, 76; Oakington 10, 11, 20, 74, 165; St Eval 20, 21; Shepherd's Grove 156; Stoney Cross 157; Stradishall 80, 117; Syerston 38; Tangmere 42; Tempsford 119, 120; Tilstock 76; Tuddenham 108, 110, 112, 115, 116; Waterbeach 86, 162, 163; Weston Zoyland 128; Wethersfield 124; Woodbridge 105; Wratting Common (West Wycombe) 108; Wyton 15, 16, 43, 161
Royal Marines 93
Royal Navy 21, 93
Royston, G. 113
Russell, Flt Sgt J. (99)

Sach, Flg Off J. 14

St Albans, Herts. 144
Salisbury, Wilts. 1
Salisbury Plain 128
Salvation Army 63, 64
Sandvik, H. 155
SAS 123, 124
Saunders, H. St G. 165
Scharnhorst 20, 21
Scherfling, Fw K-H. (21)
Seabolt, Sgt 54, 56
Searby, J. 38
SEBRO 43–53
Shopland, Flt Sgt A. 152, 156
Short, O. 5
Short Brothers
 & Harland, Belfast 12, 22, 30, 44; Rochester 1, 4, 10, 22, 30, 31, 32, 44; Swindon 30, 44, 163
Shotts, Lanarkshire 40
Shrump, WO E. 135
Slee, Flt Lt 1
Smith, Flt Lt 1
Smith, Sgt J. 64
Smith, Sgt W. 94
Smithers, Sqn Ldr 18
Songedal, P. 154
Spain 64
Spanish Civil War 65
Speare, Sqn Ldr D. 16
Speed up on Stirlings 163
Spiby, Sgt J. 80
Spitfire, Supermarine 157, 160, 168
Stalag IVb 65
Stanley, Flt Sgt A. 100
Steeple Ashton, Wilts. 142
S. 31, Short 3, 4
Stimson, J. 110
Stimson, Flt Sgt M. 136, 137
'Stirlingaires', the 49, 52, 53
Stock, G. 8, 11
Stoddard, Flt Lt 99
Studd, Plt Off R. 39, 42
Sunderland, Short 3, 29, 168
Sweden
 Stockholm 157
Switzerland
 Geneva 38
Sydenham, NI 7, 29

Target For Tonight 162, 163
Taylor, WO A. 152, 156
That Old-Fashioned Stirling of Mine 92
The Aeroplane 22, 161
The Dambusters 163
The Greatest People in the World 165

The World in Ripeness 165
Thomas, Flt Sgt 1
Thomas, G. 122
Thomlinson, Capt. 2
Thompson, T. 105, 106, 107
Thomson, Flt Sgt D. 42
Thygesen, J. 146, 148, 149, 150, 155
Towers, Flt Sgt J. 108, 109, 110, 117
Town, G. 121, 122
Trans-Air, Belgium 160
Traynor, Flg Off J. 40, 41
Trenchard, MRAF Sir H. 2
Trevor-Roper, Flt Lt R. 156
Triptree, Sqn Ldr D. 131, 134
Truelove, T. 135
Turner, A. 113, 114

Ultra Electric, Acton 32

Vaaje, K. 154
Vaaland, T. 156
Vogt, Hpt 147

Waje, P. 154
Waller, C. 112, 113, 114, 115
'Warships Week' 67
Warspite, HMS 135
Warwick, Vickers 160
Waters, J. 100
Watt, Sqn Ldr G. 91
Watt, H. 162
Webb, Flt Sgt T. 124, 144
Webster, E. 110, 118
Welch, A. 46
Wellington, Vickers xiii, (xiv), 1, 10, 22, 23, 24, 37, 75, 162
White, Sgt 14
White, Flt Sgt O. (99)
White, Sgt T. 124, 125, 144
Whitley, Armstrong Whitworth 22, 37
Wickson, Sgt H. 89
Wilkins, Plt Off L. 157
Wilson, Sgt J. 54, 56
Wiltshire News 145
'Wings for Victory' 66, 67, 68, 70, 71, 166
Witt, D. 14, 20
Wood, Flt Sgt J. 42
Woodhall, J. 67, 70
'Workers' Playtime' 49, 53
Wright, Sgt G. 93, 94, 96, 99
Wright, V. 52, 53

York, Avro xiv